RUDIMENTS OF ALGEBRAIC GEOMETRY

W. E. JENNER

DOVER PUBLICATIONS, INC.
Mineola, New York

Bibliographical Note

This Dover edition, first published in 2018, is an unabridged republication of the work originally published in 1963 by Oxford University Press, New York.

Library of Congress Cataloging-in-Publication Data

Names: Jenner, William Elliott, 1924- author.
Title: Rudiments of algebraic geometry / W.E. Jenner.
Description: Dover edition. | Mineola, New York : Dover Publications, Inc., 2018. | Originally published: New York : Oxford University Press, 1963. | Includes bibliographical references.
Identifiers: LCCN 2017029716 | ISBN 9780486818061 | ISBN 0486818063
Subjects: LCSH: Geometry, Algebraic.
Classification: LCC QA564 .J4 2018 | DDC 516–dc23
LC record available at https://lccn.loc.gov/2017029716

Manufactured in the United States by LSC Communications
81806301 2017
www.doverpublications.com

CONTENTS

INTRODUCTION

One of the most significant developments in recent mathematics is the resurgence of interest in algebraic geometry, a trend dating more or less from the publication of Weil's *Foundations*. Up to the present time, this revival has manifested itself only at the graduate level and beyond. At the undergraduate level on the other hand, the geometric traditions, if represented at all, usually have been presented in the form of a course in analytic or synthetic projective geometry, culminating in the theory of the conic sections. The transition from this to modern algebraic geometry is beset with imposing obstacles, as most people who have gone through the experience will testify. As matters stand at present, to exaggerate only slightly, students have virtually no geometric experience of much significance between their freshman study of conics and quadrics and some graduate course devoted, more likely than not, to proving the Riemann-Roch Theorem in six easy lessons — using sheaves — and this sometimes without having ever beheld a curve possessed of a singular point. One purpose of this book is to assist graduate students who find themselves in this position and to obviate for them, at least partially, the necessity of struggling with the confusion, obscurity, and downright error which sometimes arise as one extracts the needed information from some of the older literature.

Nevertheless, the book is addressed primarily to undergraduates and is intended to supplement, or to provide an alternative to, more traditional subject matter. In this way it is hoped to suggest some idea of the actual concerns of present-day geometers, and at the same time to make the way easier for those going on to a serious study of algebraic geometry. The prerequisites have been kept to an absolute minimum. I construe these to consist of elementary analytic geometry up through the conics and quadrics, the fundamentals of linear algebra (which may be studied concurrently), and a knowledge of calculus up through partial derivatives. A brief outline of the necessary algebra has been included by way of preamble. It is sufficient to consult this only as the necessity arises; the reader may begin safely with Chapter I.

There is perhaps a legitimate question, particularly among the experts, as to whether it is desirable — or indeed possible — to say anything worthwhile about algebraic geometry at the undergraduate level. My own answer is in the affirmative and this book represents the results of my attempt to deal with the question. In view of the absence of precedents, it is difficult to be convinced that one has chosen the "right" things to talk about. The choices made here are admittedly tentative, and it is hoped that further experimentation by others, more competent than myself, will lead to a more definitive result.

From a technical point of view, the principal aim of the book is to close part of the gap between elementary analytic geometry and abstract algebraic geometry along the lines, for instance, of Lang's recent book [10]. This entails a transition to a new attitude of mind both with regard to subject matter and to method. This transition is exemplified, for instance, in recasting the theory of tangents in algebraic terms. This necessitates reformulation of the required calculus in purely algebraic form, thus extirpating the notion of limit and allowing generalization to arbitrary ground fields. Another theme of constant recurrence is the necessity of working over a sufficiently large field, usually algebraically closed, in order that the geometric results may take on their most felicitous form. It is hoped that in this way some light will be shed on the reasons behind the rather strong initial algebraic assumptions that are usually made in abstract algebraic geometry.

A few words may clarify the point of view taken on certain topics. First of all, points are defined as n-tuples — not even as equivalence classes under certain admissible transformations. This is quite sufficient for the purpose at hand, and bases the theory on a very simple set-theoretical construction rather than on the invention of a new class of things called "points" which are apt to evoke nonsensical metaphysical questions. One consequence of the point of view adopted here is that a transformation always appears as something that moves points around, and not as something that "changes co-ordinates." It is, of course, important to understand the two ways of looking at these things. Usually, however, this is carefully dealt with in courses in linear algebra and it is unnecessary to discuss it here.

Most of what goes on in this book is done over arbitrary fields. This is the only real divergence from current practice in algebraic geometry where it is customary to work over a so-called universal domain. There are several reasons for this decision. In the first place, since generic points are not discussed here, there is no need for transcendentals. So far as algebraic closure is concerned, I have tried to show why it is desirable, and to do this one naturally has to start without it. Actually, a great deal of what is done in this book will work over finite fields, a fact that it seemed worthwhile to point out for the reassurance of young mathematicians who have just heard of Gödel's Theorem and expect the imminent collapse of mathematics! Whatever may be said of such an attitude on their part, it is certainly indicative of a serious concern for our subject and so should be regarded with sympathetic understanding. Certainly no harm is done; they either recover or else go on to become experts in mathematical logic.

The matter of terminology in algebraic geometry is at

present in an unsettled state, especially in view of certain recent activities in Paris. Fortunately, however, this difficulty is hardly relevant for a book at this level. The only conventions made here that call for comment are that all fields are assumed to be commutative and that the term "variety" is reserved for algebraic sets that are absolutely irreducible.

I am deeply indebted to several of my friends for their criticisms and helpful advice; in particular to Douglas Derry, William L. Hoyt, Kenneth May, and Maxwell Rosenlicht. The critical comments of A. Seidenberg on an early version of the manuscript were especially helpful to me. I also wish to thank Mrs. Florence Valentine for typing the manuscript. I am especially grateful to Ralph Spielman whose wise counsel and encouragement were instrumental in my decision to write this book.

Chapel Hill, North Carolina W.E.J.
September 1962

ALGEBRAIC PRELIMINARIES

We collect here some of the basic algebraic facts that we use. These are part of the standard equipment of all mathematicians and so our account is confined to the basic definitions. The account given by Bourbaki [3] of these matters is generally regarded as definitive. Among the best source books in English is Zariski and Samuel [18]. For linear algebra the books of Halmos [6], Hoffman and Kunze [8], and Jaeger [9] can be consulted; the last two give an extensive account of the computational aspects of determinant theory. Basic set-theoretical facts, which we assume known, are given in Halmos [5].

A set S is said to admit a *law of binary composition* if with each ordered pair (x,y) of elements of S is associated an element of S. This element will be denoted here by $x \circ y$ and will be called the "product" or "sum" of x and y (in that order), whichever is the more suggestive in a particular context.

1. A *group* is a set G together with a binary law of composition $(a,b) \longrightarrow a \circ b$ such that

(i) G is closed under ($\circ$); that is, $a \circ b$ is defined and is an element of G for all $a, b \in G$.

(ii) $a \circ (b \circ c) = (a \circ b) \circ c$ for all $a, b, c \in G$.

(iii) The equations $a \circ x = b$ and $y \circ a = b$ each have solutions for any $a, b \in G$.

Remark: For purposes of verification, axiom (iii) is usually replaced by axioms (iii′) and (iii″) together:

(iii′) There is an element $e \in G$, the *identity element*, such that $a \circ e = a = e \circ a$ for all $a \in G$.

(iii″) For every $a \in G$ there is an element $a' \in G$, the *inverse* of a, such that $a \circ a' = a' \circ a = e$.

If $a \circ b = b \circ a$, the group G is said to be *abelian* or *commutative*.

2. A *ring* is a set R together with two laws of binary composition, $(a,b) \longrightarrow a + b$ and $(a,b) \longrightarrow a \cdot b$, called addition and multiplication respectively, such that

(i) R is an abelian group under addition.

(ii) R is closed under multiplication.

(iii) $a \cdot (b \cdot c) = (a \cdot b) \cdot c$ for all $a, b, c \in R$.

(iv) $a \cdot (b + c) = a \cdot b + a \cdot c$ and $(a + b) \cdot c = a \cdot c + b \cdot c$ for all $a, b, c \in R$.

If $a \cdot b = b \cdot a$ for all $a, b \in R$, then R is said to be *commutative*.

3. A *field* k is a commutative ring for which the set of elements not equal to 0 (where 0 is the additive identity element) form a group under multiplication. (This is often, if somewhat imprecisely, described by saying that a field behaves like the set of rational (or real) numbers with respect to the basic arithmetic rules of addition, subtraction, multiplication, and division.) A field k is said to be *algebraically closed* if every polynomial equation with coefficients in k has a root in k; for example, the field of complex numbers. If $1 + \cdots + 1$ (n times) $= 0$ in a field k, where 1 is the multiplicative identity element, then the smallest such strictly positive integer n is a prime number called the *characteristic* of k; if there is no such integer n, then k is said to be of *characteristic zero*.

4. In working over arbitrary fields, the notion of *polynomial* must be refined beyond its use in elementary algebra. The trouble is that a polynomial in the elementary sense may be identically zero without its "looking like" the zero polynomial. For instance, in this sense, the polynomial $x^2 - x$ is identically zero in the field F_2 of residue-classes of integers reduced modulo 2 (cf. Chapter I). We avoid this difficulty by regarding "x" only as a symbol: something in its own right beyond the fact that we can substitute things for it. The metaphysical hiatus in this definition can be

avoided as follows: let N be the set of integers greater than or equal to 0. A polynomial in one "variable" over a field k is a mapping $P: N \longrightarrow k$ such that $P(n) = 0$ for all $n \in N$ except a finite number. The idea is that $P(n)$ is the coefficient of what used to be called x. For further details, and the generalization to polynomials in several variables, the reader is referred to Zariski and Samuel [18].

5. Matrices occur in mathematics *almost* always in connection with linear transformations. We define matrices as follows, following Chevalley [4]: let I be the set of integers $1, 2, \cdots, m$ and J the set of integers $1, 2, \cdots, n$. An m by n matrix Φ with coefficients in a ring R is a mapping $\Phi: (i,j) \longrightarrow \alpha_{ij}$ of the cartesian product $I \times J$ into R. We can identify (for notational purposes) the matrix Φ with the array

$$\begin{pmatrix} \alpha_{11} & \cdots & \alpha_{1n} \\ \cdot & & \cdot \\ \cdot & & \cdot \\ \cdot & & \cdot \\ \alpha_{m1} & \cdots & \alpha_{mn} \end{pmatrix}.$$

The first subscript i of α_{ij} is called the *row index* and the second subscript j the *column index*. If $\Psi: (i,j) \longrightarrow \beta_{ij}$ is another such matrix, we define their *sum* to be the matrix $\Phi + \Psi: (i,j) \longrightarrow \alpha_{ij} + \beta_{ij}$. The set of m by n matrices with coefficients in R forms an abelian group under addition. If Φ is an l by m matrix and Ψ an m by n matrix, we define their product to be the l by n matrix $\Phi \circ \Psi: (i,j) \longrightarrow$ $\sum_{k=1}^{m} \alpha_{ik}\beta_{kj}$. The product $\Phi \circ \Psi$ is defined if and only if the column index of Φ is equal to the row index of Ψ. The motivation for these definitions lies in the theory of linear transformations. The set $[R]_n$ of n by n matrices with coefficients in a ring R itself forms a ring under the matrix operations.

6. A *vector space* V over a field k is an additive abelian group for which a multiplication of elements of V by ele-

ments of k is defined such that

(i) $\lambda(u + v) = \lambda u + \lambda v$.

(ii) $(\lambda + \mu)v = \lambda v + \mu v$.

(iii) $\lambda u = u\lambda$.

(iv) $1\, u = u$.

Here u, $v \in V$ and λ, $\mu \in k$ and 1 is the multiplicative identity element of k. (In some parts of mathematics, axiom (iii) is weakened so that there is a distinction between left and right vector spaces. This generalization will be unnecessary for our purposes.)

A particularly important example is the vector space of n-tuples over a field k. If $(x_1, \cdots, x_n)$ and $(y_1, \cdots, y_n)$ with $x_i, y_i \in k$ are two such n-tuples we define addition component-wise, $(x_1, \cdots, x_n) + (y_1, \cdots, y_n) = (x_1 + y_1, \cdots, x_n + y_n)$, and define $\lambda(x_1, \cdots, x_n) = (x_1, \cdots, x_n)\lambda = (\lambda x_1, \cdots, \lambda x_n)$ for $\lambda \in k$. (For readers of austere tastes, n-tuples over k can be defined as mappings of the set of integers $1, 2, \cdots, n$ into k. An even more austere approach, in terms of the most primitive set-theoretical notions is given in Halmos [5].)

A set of vectors $v_1, \cdots, v_n \in V$ is said to be *linearly dependent* if there is a relation $\lambda_1 v_1 + \cdots + \lambda_n v_n = 0$ with $\lambda_i \in k$ and some $\lambda_i \neq 0$; otherwise it is *linearly independent.* A vector space V is said to be *finite dimensional* if there exist a finite number of vectors $v_1, \cdots, v_n$ such that every $v \in V$ can be expressed in the form $v = \lambda_1 v_1 + \cdots + \lambda_n v_n$ with the $\lambda_i \in k$. If this is so, then the vectors $v_1, \cdots, v_n$ can be chosen to be linearly independent; then the λ_i are unique, and the integer n is unique. This value of n is called the *dimension* of V and $v_1, \cdots, v_n$ constitute a *basis* of V. Any n linearly independent elements of V form a basis.

7. Let V be a vector space of dimension n over k. A *linear transformation* of V is a mapping of V into V, having the properties $\phi(u + v) = \phi(u) + \phi(v)$ and $\phi(\lambda u) = \lambda\phi(u)$ where u, $v \in V$, and $\lambda \in k$. If ψ is another linear transformation of

V we define $\phi + \psi : u \longrightarrow \phi(u) + \psi(u)$ and $\phi \circ \psi : u \longrightarrow \phi(\psi(u))$. The linear transformations of V form a ring. We say ϕ is *invertible* if there is a linear transformation ϕ' such that $\phi \circ \phi' = \phi' \circ \phi$ is the identity mapping on V. The elementary theory of determinants will be assumed known. Determinants are defined directly for linear transformations in Halmos [6]. A linear transformation is invertible if and only if its determinant is non-zero. For an account of the techniques of determinant theory see Jaeger [9].

I

AFFINE SPACES

1. Affine spaces and algebraic sets

Algebraic geometry is concerned with the study of loci of polynomial equations. For the most part, metrical properties of euclidean geometry, distance, and angle, are not considered, and so we work in rather more general kinds of spaces called *affine* and *projective*. These can be defined synthetically along the lines of Euclid's *Elements*, but it is customary nowadays in algebraic geometry to start with purely algebraic definitions.

In this chapter we shall consider affine spaces. Let k be any field. By *affine n-space over k*, denoted $A_n(k)$, is meant the set of n-tuples $(x_1, \cdots, x_n)$ where $x_i \in k$. Such an n-tuple is called a *point*. In case k is the field of real numbers, it is harmless to think of $A_n(k)$ as being euclidean n-space without any mention of distance. In particular, if $n = 2$ or 3 it is possible to draw "pictures" (graphs) in the usual sense. If $f(x_1, \cdots, x_n) = 0$ is a polynomial equation in n variables with coefficients in k, the set of points in $A_n(k)$ satisfying this equation is called an *algebraic hypersurface*. Since we shall consider only loci given by polynomial equations, the adjective "algebraic" will generally be omitted. The intersection of a finite number of hypersurfaces is called an *algebraic set*. (Some writers use the term "algebraic variety"; a convention seems to have developed, however, to reserve this for algebraic sets with certain extra conditions which are too technical for discussion here.) If $n = 2$ or 3, hypersurfaces are called *curves* or *surfaces*, respectively. A hypersurface given by an equation of degree 1 is called a *hyperplane* and in case $n = 2$ or 3, these are called *lines* or *planes*, respectively. An algebraic set in $A_n(k)$ is said to be *reducible* if it can be expressed as the union of two proper algebraic subsets; otherwise it is *irreducible*. For instance, the curve given by $(x_2 - x_1)(x_2 - x_1^2) = 0$ in $A_2(R)$, where R is the field of real numbers, is the union of the line $x_2 = x_1$ and the parabola $x_2 = x_1^2$. Reducibility properties depend strongly on the nature of the field. Consider, for instance, the equation $x_1^2 + x_2^2 = 0$. Over $A_2(R)$ the locus is a single point, but over $A_2(C)$, where C is the field of complex numbers, the locus consists of two lines $x_1 + ix_2 = 0$ and $x_1 - ix_2 = 0$. To treat the notion of reducibility in its fullest generality it has been found best to take k algebraically closed, even though the coefficients in the defining equations of the locus may be from a quite "small" field. Actually, this convention is made almost universally in present-day algebraic geometry. It will be not made in this book, for the simple reason that we do not go far enough for it to make an important difference. Indeed, by *not* making this assumption, we shall be in a better position to show more clearly, in some simple situations, why it is desirable. Also, what is perhaps more important at this stage, we very often wish to work over the real numbers. This has the enormous advantage that it is possible to draw pictures.

So far as we have gone, the field k may even be taken finite. Let us recall the construction of the simplest such fields. Let Z be the ring of positive and negative integers and let p be a fixed prime number. For $a, b \in Z$ we say a is congruent to b modulo p if $a - b$ is divisible by p. This is written $a \equiv b \pmod{p}$. This relation induces a partition of Z into disjoint subsets, called residue-classes. A residue-class consists of the set of all integers congruent to a given one. The residue-class containing an integer a is denoted by $[a]$. We define addition and multiplication of residue classes by $[a] + [b] = [a + b]$ and $[a] \cdot [b] = [a \cdot b]$, respectively. (It has to be shown, of course, that these definitions depend only on the residue-classes and not on the choice of their representatives.) The set of residue-classes with these operations forms a field with p elements. This field is denoted by F_p. (If p is not a prime, the preceding construction still works, but the resulting number system is not a field, even though it is still a commutative ring.) It is customary to denote the elements of F_p by representatives for the residue-classes and identify them if their difference is divisible by p. In this sense F_3 consists of the integers 0, 1, 2. Then we would write in this field 3 "=" 0, 4 "=" 1, etc. The space $A_2(F_3)$ has nine points which can be represented as follows:

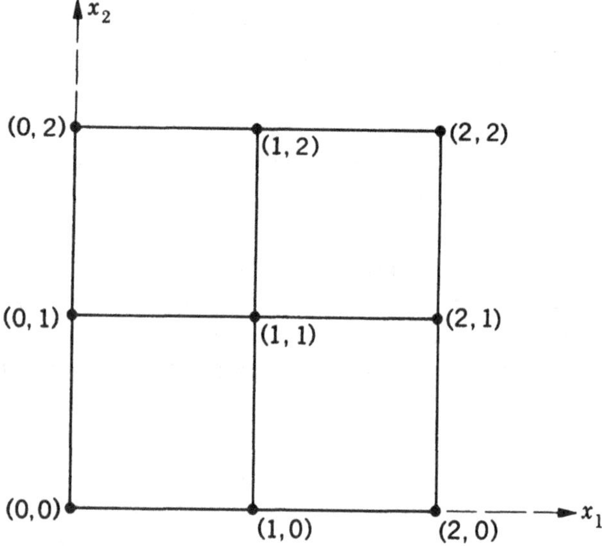

FIG. 1.1

The points in $A_2(F_3)$ in this diagram are marked by the heavy dots. Note that (3,2) is the same point as (0,2), for instance. It sometimes is helpful to think of this diagram as being drawn on an elastic sheet which is deformed as follows: the line that ordinarily in euclidean geometry would have the equation $x_2 = 3$ is pulled down and attached along the x_1-axis so that a point $(a, 3)$ is attached to the point $(a, 0)$ in the sense of ordinary euclidean co-ordinates. Then the line $x_1 = 3$ (which is now a circle) is pulled over to match up with the line $x_1 = 0$ so that a point $(3, a)$ is attached to the point $(0, a)$. The result is that we now have the nine points of $A_2(F_3)$ represented by nine points on a torus (a doughnut-shaped surface) in ordinary euclidean 3-space. Whenever some "peculiar" space is represented by suitable objects in ordinary euclidean space, such a representation is called a *euclidean model* of that space. The construction of such models has two purposes: first, to remove psychological qualms about the new space, and second, to show that if euclidean space is "all right" so is the new space. This latter point is of extreme logical importance. Such a method is used, for instance, to demonstrate the consistency—relative to euclidean geometry—of the classical non-euclidean geometries. In the next chapter the same idea comes up when we construct a euclidean model for the real projective plane.

To return to $A_2(F_3)$, it must be understood that the only points in the whole space are the nine points marked on the diagram. There are no points "in between"! In this space we can talk about curves as we do in more familiar spaces. For instance, consider the equation $x_2 = x_1^2$. We just construct a table of values as usual:

x_1	0	1	2
x_2	0	1	1

The graph—the whole curve—consists of the three points marked by the heavy dots as follows:

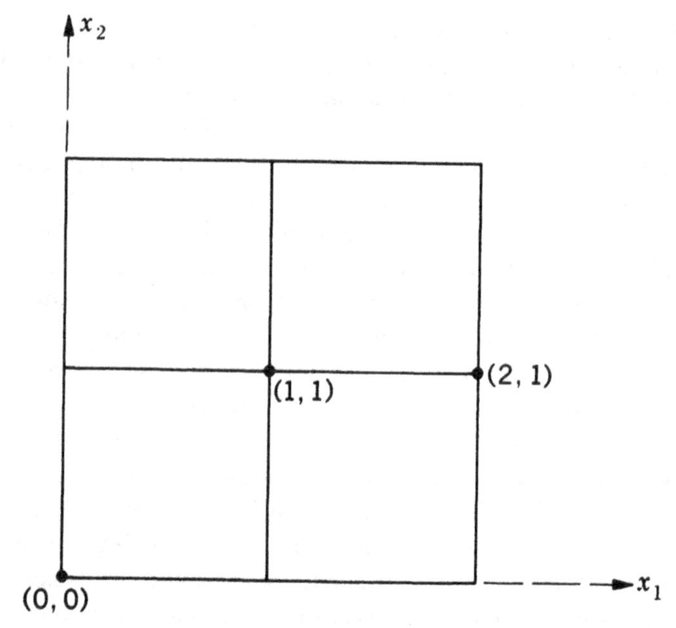

FIG. 1.2

2. Straight lines and hyperplanes

In elementary analytic geometry many important properties of curves and surfaces are studied by intersecting them by lines or planes respectively. The matter of tangents comes under this category. Also, the shape of the quadric surfaces in euclidean space is determined by slicing them with planes and looking at the curves of intersection. The utility of first-degree loci for studying geometrical objects still persists in general affine spaces, so it will be necessary to look at them a little more closely. We shall first consider lines in $A_2(k)$.

Suppose $ax + by + c = 0$ is the equation of a straight line in $A_2(k)$. Then not both a and b are zero. We wish to obtain a parametrization of this line. Assume that $a \neq 0$ and that (x_0, y_0) is a point on the line. (It is trivial to see there must be at least one such point.) Then $ax_0 +$

$by_0 + c = 0$ and so a point (x, y) is on the line if and only if

$$(1) \qquad a(x - x_0) + b(y - y_0) = 0.$$

Set

$$t = \frac{y_0 - y_0}{a}.$$

Then

$$(2) \qquad \begin{cases} x = x_0 + bt, \\ y = y_0 - at. \end{cases}$$

These equations give parametric equations for the line since, if t is assigned arbitrarily in k, the point (x, y) given by (2) will lie on the line. The same result is obtained if we start from the assumption $b \neq 0$. It is easy to verify that, if the line can also be parametrized in the form

$$\begin{cases} x = x_0 + b's, \\ y = y_0 - a's, \end{cases}$$

then there is a non-zero element $c \in k$ such that $a' = ca$ and $b' = cb$.

Remark: In elementary analytic geometry of the euclidean plane, a straight line is ultimately defined in terms of the intuitively acceptable characterization that for any two points (x_1, y_1) and (x_2, y_2) on the line, the ratio of $x_1 - x_2$ and $y_1 - y_2$ is a constant. (Unfortunately this fact is seldom made sufficiently explicit in elementary text books.) In view of this fact, the equations (2) above should settle the conscience of anyone who is disturbed by *defining* a line to be the locus of a first-degree polynomial equation.

A parametrization similar to (2) can be given in the higher-dimensional cases of planes and hyperplanes. This can be done best by using techniques of linear algebra. Here, however, we prefer to adopt a more naïve procedure which, although less elegant, is quite sufficient for our purposes. Suppose a hyperplane is given by the equation $a_1x_1 + \cdots + a_nx_n + b = 0$. Suppose $a_1 \neq 0$ and that $(c_1, \cdots, c_n)$ is a point on the hyperplane. Then $a_1(x_1 - c_1) + \cdots + a_n(x_n - c_n) = 0$. For $i = 2, \cdots, n$ set $t_i = \dfrac{c_i - x_i}{a_1}$. Then for any point $(x_i, \cdots, x_n)$ on the hyperplane

$$(3) \qquad \begin{cases} x_1 = c_1 + a_2t_2 + \cdots + a_nt_n \\ x_2 = c_2 - a_1t_2 \\ \cdot \\ \cdot \\ \cdot \\ x_n = c_n - a_1t_n \end{cases}$$

Conversely, if $t_2, \cdots, t_n$ are assigned arbitrarily in k, then the point $(x_1, \cdots, x_n)$ given by (3) will lie on the hyperplane and so equations (3) constitute a set of parametric equations for the hyperplane. If $a_1 = 0$, the procedure above must be modified by using in place of a_1 some other coefficient which is not zero.

3. Intersections of lines with algebraic curves

In analytic euclidean geometry it is customary to study the problem of tangents by means of the calculus. If attention is restricted to algebraic curves it would be desirable to treat this problem in a purely algebraic manner—without using the notion of limit which lies at the foundation of the calculus. If we wish to work over more general fields than the real numbers, the absence, in general, of any notion of limit forces us into such a procedure. To this end we now formalize, in a purely algebraic way, those parts of the calculus that are necessary for our purpose.

We start with the ring $k[X]$ of polynomials in a variable X with coefficients in a field k. The derivation operator D_X on $k[X]$ is defined purely algebraically by

$$D_X\left(\sum_{k=0}^{n} a_k X^k\right) = \sum_{k=1}^{n} a_k \cdot k X^{k-1}.$$

Using this definition, one can prove directly all the usual rules about derivatives of sums and products. If $f(X) \in k[X]$ we as usual denote the n-th derivative of $f(X)$ by $f^{(n)}(X)$.

If $f(X) = \sum_{k=1}^{n} a_k X^k$

then

$$f(X) = f(0) + f'(0) \cdot X + \frac{f''(0)}{2!} X^2 + \cdots + \frac{f^{(n)}(0)}{n!} X^n$$

provided the characteristic of the field k is greater than n (so we can divide by $n!$). This statement is verified directly by computing successive derivatives at $X = 0$ and showing that $a_k = \dfrac{f^{(k)}(0)}{k!}$. This expansion for $f(X)$ is called the formal Maclaurin series for $f(X)$. (In case k is the field of real numbers, it is the same as the Maclaurin series in the usual sense which in this case terminates.) For the ring $k[X_1, \cdots, X_n]$ of polynomials in n variables $X_1, \cdots, X_n$ with coefficients in k, the partial derivative operators $\dfrac{\partial}{\partial X_i}$ are defined purely algebraically in the obvious manner. If $X_i = g_i(t)$ where t is an independent variable and $g_i(t)$ is a polynomial in t, then

$$D_t f(X_1, \cdots, X_n) = \sum_{i=1}^{n} \frac{\partial f}{\partial X_i} \cdot D_t X_i$$

where $f(X_1, \cdots, X_n) \in k[X_1, \cdots, X_n]$. With this machinery we may now proceed to our main task.

Let C be a curve in $A_2(k)$ given by a polynomial equation $f(x, y) = 0$ of degree n. It will be assumed temporarily that k has characteristic either zero or greater than n. Let $p = (x_0, y_0)$ be an arbitrary point on C and consider a line L through this point. The parametric equations of L are of the form $\begin{cases} x = x_0 + at \\ y = y_0 + bt \end{cases}$ with not both a and b being zero. The intersections of L and C are given by the solutions in t to the equation $f(x_0 + at, y_0 + bt) = 0$. Set $F(t) = f(x_0 + at, y_0 + bt)$ and expand this in a formal Maclaurin series

$$F(t) = F(0) + \frac{F'(0)}{1!}t + \cdots + \frac{F^{(n)}(0)}{n!}t^n.$$

We say that *L cuts C at (x_0, y_0) with multiplicity m* if $F^{(m)}(0) \neq 0$ and $F^{(l)}(0) = 0$ for $l < m$. It is easy to show that this definition is independent of the particular parametrization of the line L. (The proof depends on the fact that the parameters a and b are determined to within a non-zero proportionality factor.)

Now suppose this intersection multiplicity is greater than or equal to 2. Then $F'(0) = 0$. This means that

$$\left(\frac{\partial f}{\partial x}\right)_p a + \left(\frac{\partial f}{\partial y}\right)_p b = 0. \tag{1}$$

(The subscripts mean that the derivatives are evaluated at (x_0, y_0).) The point (x_0, y_0) is said to be *singular* if

$$\left(\frac{\partial f}{\partial x}\right)_p = \left(\frac{\partial f}{\partial y}\right)_p = 0;$$

otherwise it is *simple*.

Suppose that (x_0, y_0) is a simple point. Then the equation (1) determines the ratio of a and b and so determines the line L uniquely. This can be seen as follows: suppose

$\left(\frac{\partial f}{\partial x}\right)_p \neq 0$ and take $b = 1$. Then $a = -\frac{\left(\frac{\partial f}{\partial y}\right)_p}{\left(\frac{\partial f}{\partial x}\right)_p}$. The

equation of L in non-parametric form is $b(x - x_0) - a(y - y_0) = 0$ and so the equation of the line cutting C at (x_0, y_0) with multiplicity greater than or equal to 2 is

$$\left(\frac{\partial f}{\partial x}\right)_p (x - x_0) + \left(\frac{\partial f}{\partial y}\right)_p (y - y_0) = 0. \tag{2}$$

This line is called the *tangent* to C at (x_0, y_0). We end up with exactly the same equation if we start with the assumption $\left(\frac{\partial f}{\partial y}\right)_p \neq 0$.

A slightly different way to arrive at equation (2) is this: suppose that (1) is satisfied and that (x_1, y_1) is a point on L cutting C at p with multiplicity greater than or equal to 2. Then

$$\left(\frac{\partial f}{\partial x}\right)_p at + \left(\frac{\partial f}{\partial y}\right)_p bt = 0$$

for all $t \in k$ and so (x_1, y_1) is on the line (2). Conversely, suppose (x_1, y_1) is on the line (2). Set $a' = x_1 - x_0$ and $b' = y_1 - y_0$. Then

$$\left(\frac{\partial f}{\partial x}\right)_p a' + \left(\frac{\partial f}{\partial y}\right)_p b' = 0$$

and so (x_1, y_1) is on a line cutting C at p with multiplicity greater than or equal to 2. Thus (2) is the equation of the unique line having this property.

4. Singular points

Suppose (x_0, y_0) is a singular point of the curve so that $\left(\frac{\partial f}{\partial x}\right)_P = \left(\frac{\partial f}{\partial y}\right)_P = 0$. Then every line through this point cuts the curve with multiplicity at least 2 at that point. Suppose we want L to cut C at (x_0, y_0) with multiplicity at least 3. The condition for this is $F''(0) = 0$ which means

$$(1) \qquad \left(\frac{\partial^2 f}{\partial x^2}\right)_P a^2 + 2\left(\frac{\partial^2 f}{\partial x \partial y}\right)_P ab + \left(\frac{\partial^2 f}{\partial y^2}\right)_P b^2 = 0.$$

If not all second partials vanish at the singular point (x_0, y_0), it is called a *double point*. In general, (x_0, y_0) is called an *r-fold singular point* if all partials of order strictly less than r vanish there, but some r-th partial does not.

Let us assume now that (x_0, y_0) is a double point. Then (1) is a quadratic equation for the ratio of a and b. What happens now depends very strongly on properties of the field k. This is different from the situation with tangents at a simple point. The reason is this: in the case of a simple point the equation determining a and b for the tangent is linear and linear equations work perfectly in all fields; however, for a singular point the equation for a and b is at least quadratic, and such equations behave very differently in different fields.

By using the same kind of argument as the second one we used to get the tangent at a simple point (end of §3), it is easy to show that a point (x, y) lies on a line L cutting C at (x_0, y_0) with multiplicity greater than or equal to 3 if and only if

$$(2) \qquad \left(\frac{\partial^2 f}{\partial x^2}\right)_P (x - x_0)^2 + 2\left(\frac{\partial^2 f}{\partial x \partial y}\right)_P (x - x_0)(y - y_0) + \left(\frac{\partial^2 f}{\partial y^2}\right)_P (y - y_0)^2 = 0.$$

(Note that our assumption about the characteristic of k implies that this equation cannot reduce to an identity.)

There are three cases to distinguish. If the locus of (2) consists of two straight lines, the point (x_0, y_0) is called a *node*. If the locus consists of only one straight line, then (x_0, y_0) is called a *cusp*. The remaining case is that in which the locus of (2) consists of the single point (x_0, y_0), in which case the point is said to be *isolated*. (It follows from results in §6 of Chapter II that this exhausts the possibilities; that is, the locus consists of two straight lines at most.) In the cases in which lines occur, these lines are called the tangents at (x_0, y_0).

It is probably high time we defined the notion of tangent explicitly. Suppose p is a point on a curve C. Then there is an integer r such that every line through p cuts the curve there with multiplicity greater than or equal to r, but not all such lines cut with multiplicity $r + 1$. Any line that *does* cut the curve at C at p with multiplicity at least $r + 1$ is called a *tangent* at p to C.

The preceding discussion of double points is easily generalized to r-fold singular points. The equation analogous to (2) giving the tangent lines at (x_0, y_0), or this single point in case it is isolated, is

$$(3) \quad \left(\frac{\partial^r f}{\partial x^r}\right)_p (x - x_0)^r + \binom{r}{1}\left(\frac{\partial^r f}{\partial x^{r-1}\partial y}\right)_p (x - x_0)^{r-1}(y - y_0) + \cdots + \binom{r}{r}\left(\frac{\partial^r f}{\partial y^r}\right)_p (y - y_0)^r = 0.$$

When k is the field of real numbers, a curve in $A_2(k)$ can be interpreted as a curve in the ordinary euclidean plane, and so we can draw a picture of it. We now consider a few examples to illustrate some of the more elementary types of singularities.

Example 1: $y^2 = x^3 + x^2$

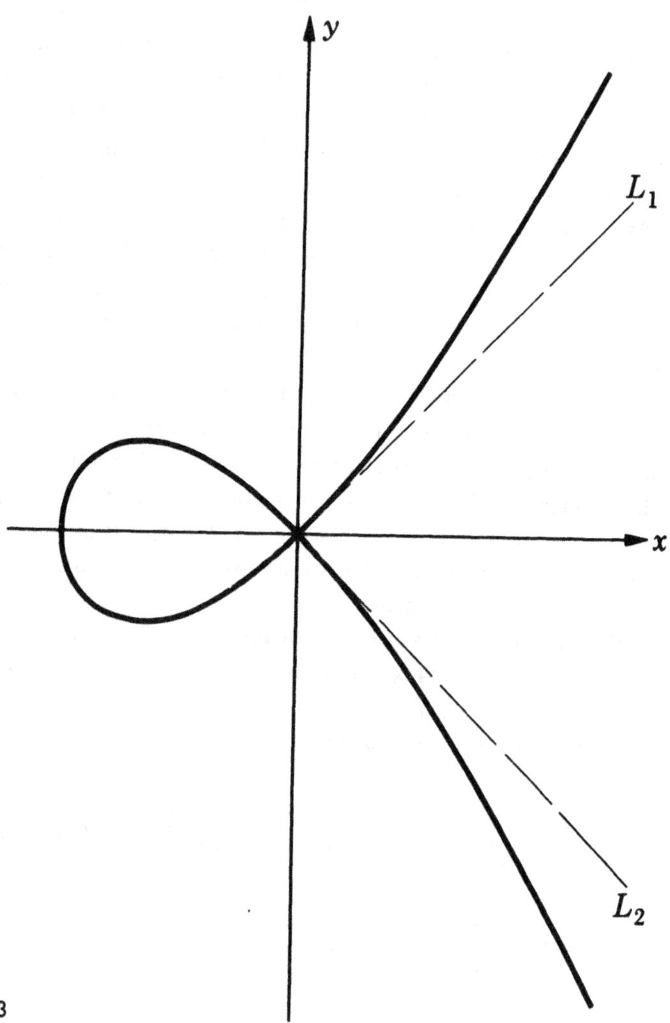

FIG. 1.3

Here there is a double point at the origin and equation (2) yields two tangent lines L_1: $y = x$ and L_2: $y = -x$. The origin is a *node*.

Example 2: $y^2 = x^3$

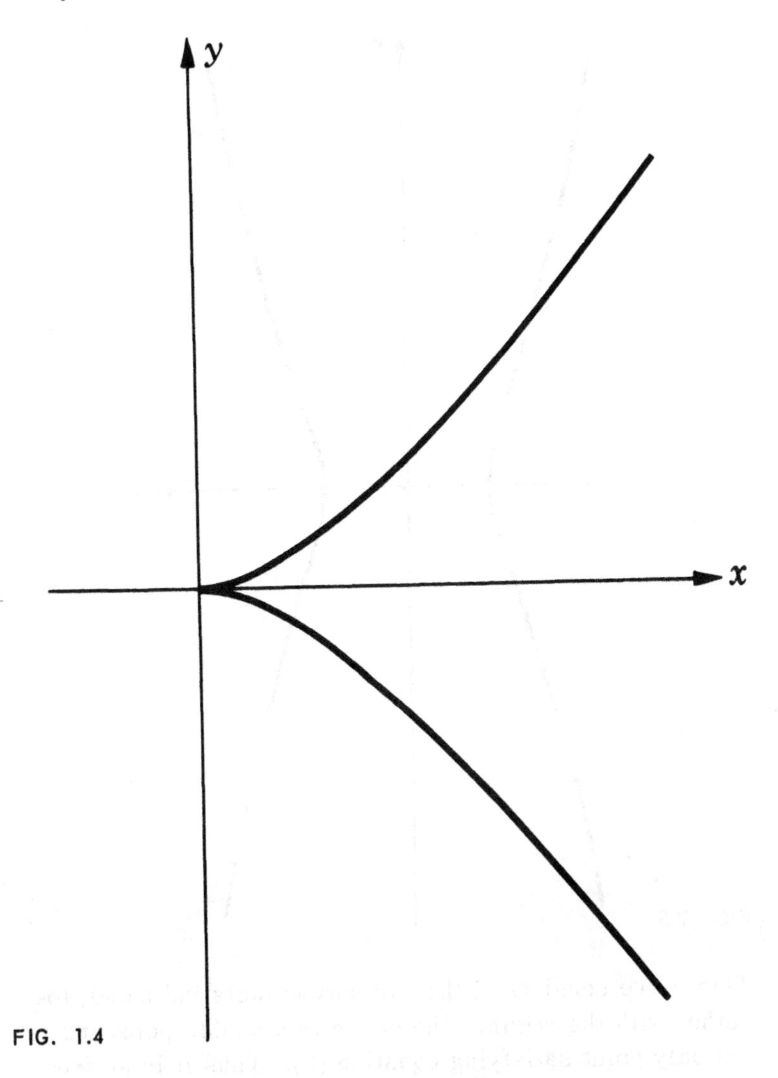

FIG. 1.4

Here again there is a double point at the origin, but equation (2) yields only one line — the x-axis. The origin is a *cusp*.

Example 3: $y^2 = x^4 - x^2$

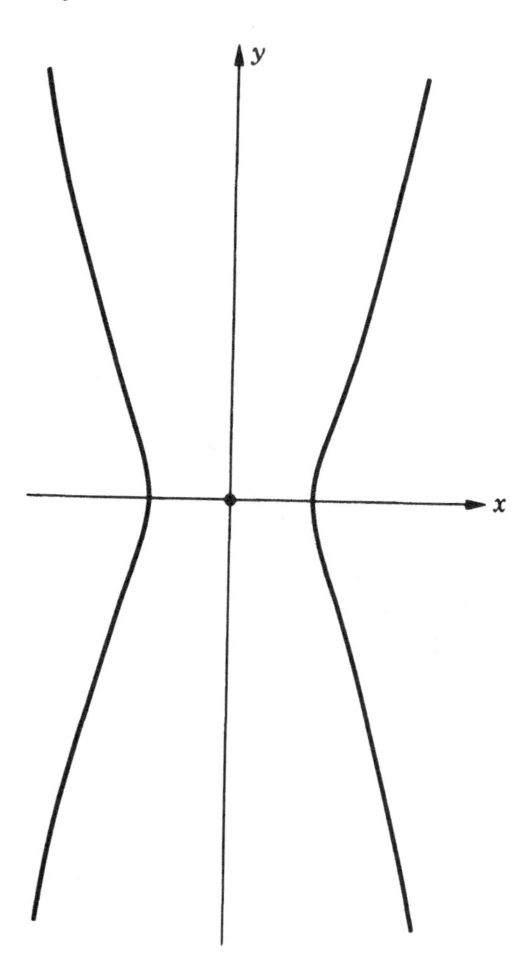

FIG. 1.5

This curve consists of the two curved parts indicated, together with the origin. The origin is a double point and is the only point satisfying equation (2). Thus it is an *isolated* point in the sense previously defined. Notice that this point is isolated from the rest of the curve in the nontechnical intuitive sense also (and in the topological sense as well).

Example 4: $x^6 - x^2y^3 - y^5 = 0$

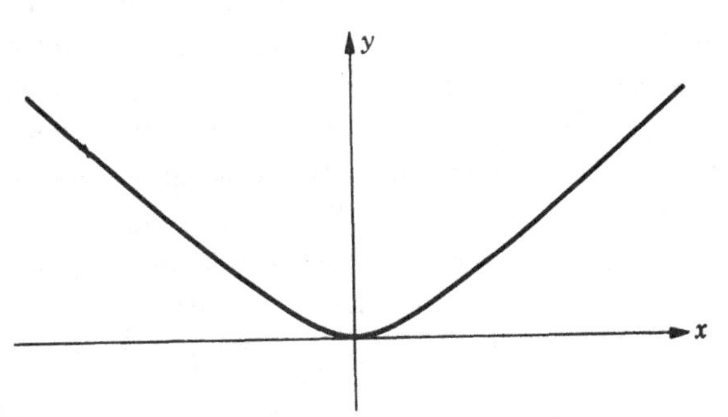

FIG. 1.6

For this curve it seems only fair to reveal that the graph is obtained from a rational parametrization; see example 2 of Chapter III. The origin is a 5-fold singular point and equation (3) reduces to $y^3(x^2 + y^2) = 0$. Over the real numbers there is only one solution, $y = 0$. This is the equation of the tangent at the origin. Over the complex numbers there would, of course, be two other tangents there. The remarkable feature of this example is that the real curve is quite smooth at the origin; there is nothing about the behavior of the real tangent there that is at all "singular" in the sense of Sherlock Holmes.

5. Modifications for the case of fields of "small" characteristic

We now add a few remarks to take care of the case in which the field has a prime characteristic p less than the degree of the curve. Keeping the same notation as in §3, we write $F(t) = c_0 + c_1t + \cdots + c_n t^n$ as a polynomial in t. Then L cuts C at p with multiplicity m if $c_m \neq 0$ and $c_l = 0$ for $l < m$. This definition still makes sense — independent of the characteristic of k. The only difference is that we cannot express a coefficient c_l in the form $\dfrac{F^{(l)}(0)}{l!}$ if the characteristic of k is less than or equal to l. Note that the c_l are homogeneous polynomials in a and b. Thus, if tangents are defined as before, then their parameters a and b are obtained by solving an equation of the form $c_l = 0$. The point p will be called *singular* if all lines through p cut the curve there with multiplicity greater than or equal to 2; otherwise it is *simple*. Thus p is singular if and only if c_1 is identically zero as a (linear) function of a and b. Now $c_1 = F'(0)$ and so $c_1 = \left(\frac{\partial f}{\partial x}\right)_p a + \left(\frac{\partial f}{\partial y}\right)_p b$. Note that in this case there is no trouble from the characteristic of k. It follows that p is singular in the sense just defined if and only if $\left(\frac{\partial f}{\partial x}\right)_p = \left(\frac{\partial f}{\partial y}\right)_p = 0$. In case p is simple, then the tangent is still given by equation (2) of §3.

To summarize the situation briefly, the previous techniques work until we have to consider some coefficient c_l where l is greater than or equal to the characteristic of k.

6. The problem of pictorial representation

The examples we have considered in §4 bring up a rather delicate question. In the theory we have developed so far, the field k has been arbitrary and so there is in general no possibility of drawing a graph of the curve in the usual sense. However, in case k is the field of real numbers, we can draw graphs. The question then arises whether the concepts we have introduced so far in a purely algebraic way — particularly the idea of tangents — coincide with the intuition we have about these things. This kind of problem arises frequently in mathematics when an attempt is made to formulate rigorously within some preassigned framework — in our case algebraic — some kind of intuitive concept. Very often we get more than we bargained for. To mention just one well-known instance, we might recall the attempts made to define a continuous curve. One of the earliest serious attempts was to define such curves as those given in parametric form $x = f(t)$, $y = g(t)$, where $f(t)$, $g(t)$ are continuous functions on the unit interval. The thing that was not bargained for in this case was brought to light by the Italian mathematician Peano. He gave an example of a curve in this sense which went through every point in a square. This obviously does not jibe with our intuition of what a curve "ought" to be. Eventually a definition that seems to do the "right" thing was produced, but that is another story.

In our own case, the theory of tangents applies to such things as the curve $y = x^3$ in $A_2(F_5)$. This is rather startling, and is certainly a blank as far as intuition is concerned. However, the theory is quite coherent and makes algebraic sense even though its significance, to be completely honest about it, is not immediately apparent. The point we are trying to make is that such things as tangents to curves over finite fields bear a relation to the naïve conception of tangents which is analogous to the relation that Peano's space-filling curves bear to the naïve conception of continuous curves. Also, it is important to notice that in each of these situations the "extras" we obtain — beyond what was expected — turn out to be of mathematical interest.

There still remains the specific question that precipitated these considerations: Does our definition of tangent coincide with the naïve conception of tangent for algebraic curves in the euclidean plane? To deal with this question in a precise manner it is necessary to use some rather substantial theorems from the theory of functions. Instead of this however, we shall discuss the matter informally, although in such a way that a person familiar with the requisite function-theoretical apparatus can expand our proof into one that is precise.

The naïve conception of tangent is essentially as follows. Suppose C is a smooth curve in the plane and that p is a point on C. Then the tangent at p is a line through p with the following property: in some sufficiently small neighborhood of p, the line cuts the curve only at this point, but if the line is rotated ever so slightly around p as the pivot point, then a new intersection appears "very close" to p.

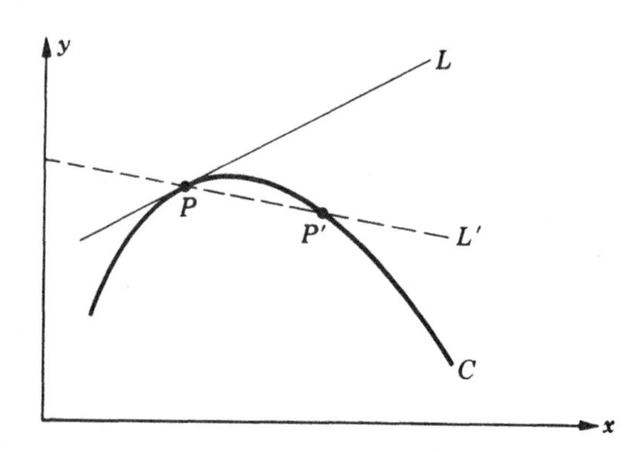

FIG. 1.7

In Fig. 1-7, the line L is the tangent at p. If it is rotated about p to the position L', a new intersection point p' develops.

Now from the algebraic point of view, the intersections of L and C are given by the solutions to the equation $f(x_0 + at, y_0 + bt) = 0$. Suppose a and b are chosen so that L is a tangent at the point $p = (x_0, y_0)$. Then this equation has a certain degree, say m, in the unknown t. The equation has a root $t = 0$ with a certain multiplicity in the ordinary sense of the theory of equations. If we count roots with their multiplicities we find there will be m of them if we work over the complex numbers. Now suppose we change the ratio of a and b by a very small amount. If this change is sufficiently small, the degree of the equation for t will not change and the multiplicity of the root $t = 0$ must strictly decrease. Thus a new intersection must arise. It is intuititively clear that there will be a new intersection "close" to p, and in the case where the curve behaves "reasonably," as in Fig. 1-7, there will be a *real* intersection p' close to p. Thus our algebraic definition seems to be in accord with the intuitive idea. This argument also applies to tangents at singular points. Actually, so far as simple points are concerned, the equation for the tangent in the algebraic sense coincides with the equation obtained in elementary calculus — a fact which is reassuring even if perhaps not very illuminating in the present context.

We have shown that if a real curve has a tangent in the naïve sense, then our algebraic theory gives its equation. On the other hand, even a real curve may have a tangent in the algebraic sense but not in the naïve sense. For instance, the curve $x^4 + y^4 + x^2 = 0$ has only the one point (0, 0) over the real numbers. According to the algebraic theory, however, this point is a cusp, and the tangent there has the equation $x = 0$. This rather bizarre situation becomes decidedly less alarming if we work over the complex numbers, since then the curve has many points. On the other hand, this requires extensive changes in the manner of representing the curve pictorially.* Nevertheless, one cannot help but feel that in the present instance the "right" thing to do is to work over the complex numbers. We shall see several examples later on, particularly in Chapter III, where we are driven to the same conclusion. The relevant property in each case is, of course, the algebraic closure of the complex numbers.

It is partially on the basis of such "experimental evidence" as we have just mentioned that the algebraic closure of the ground field is usually assumed in modern algebraic geometry. It is customary to make this assumption even when the primary interest is in what happens over

*If the equation of a complex curve is split into its real and imaginary parts, we obtain two equations in four real variables which define a surface in real affine 4-space. It can be shown that the tangent at a simple point on the complex curve then gives the corresponding tangent plane to the real surface.

a comparatively small subfield, in which case the desired results are often obtained by various "cutting-down" procedures. However, the assumption of algebraic closure along with many other basic conventions of algebraic geometry have been, and still are, subject to periodic review. For this reason they should not be regarded as final.

Irrespective of what decisions may ultimately be made on these matters, it is reasonably clear that there still will remain a question that often is disturbing to those encountering the subject for the first time, namely: is algebraic geometry, especially when done over arbitrary fields, really geometry? The uneasiness arises from the fact that algebraic geometers discuss objects for which there is often no possibility whatever of visual representation. The answer to the question depends, naturally, on what the words mean. Nowadays most mathematicians would define geometry as any part of mathematics in which utterances are made which *sound* like geometry. If this answer is open to objection on logical grounds, which manifestly is the case, the same is also true of the question. Nevertheless, the intent of this "definition" is certainly clear enough and comes about as close to the right spirit as one could hope for.

Finally, to put the matter of pictorial representation in its proper perspective, there seem to be at least two reasons for its importance. In the first place, there is no doubt that pictures of curves and other geometric objects are suggestive of appropriate geometric ideas which often generalize to the case of arbitrary fields. The second reason is much less austere but is, nevertheless, nothing to be ashamed of. This is the undeniable fact that the pictures are interesting and pleasant just to look at. The purpose of algebra definitely is not to take all the joy out of life.

Exercises for Chapter I

1. Show that if k is algebraically closed then any hypersurface in $A_n(k)$ has at least one point. Show by ex-

ample that the corresponding statement is false if k is the field of real numbers.

2. Plot the curves $y = x^3$ and $2x + 3y + 1 = 0$ in $A_2(F_5)$.
3. Supply proofs for the statements made about derivative operators at the beginning of §3. Also, show that if $f(x, y)$ is a polynomial in two variables then

$$\frac{\partial^2 f}{\partial x \partial y} = \frac{\partial^2 f}{\partial y \partial x}.$$

4. Suppose $f(x, y) = 0$ is an algebraic curve of degree n in $A_2(k)$ where k has characteristic greater than n. Suppose the point $(0, 0)$ is on the curve so that $f(x, y) = \phi_1(x, y) + \cdots + \phi_n(x, y)$ where $\phi_t(x, y)$ is homogeneous of degree t. Show that $(0, 0)$ is a singular point if and only if $\phi_1(x, y) = 0$ identically and that, otherwise, the tangent at $(0, 0)$ is $\phi_1(x, y) = 0$.
5. Notation as in exercise 4. Suppose $\phi_1, \cdots, \phi_{r-1}$ are identically zero and ϕ_r is not. Show that $(0, 0)$ is an r-fold singular point and that $\phi_r(x, y) = 0$ gives the equations of the tangents there, or the equation of the single point $(0, 0)$ in case it is isolated.
6. Suppose a hypersurface in $A_n(k)$ is given by an equation $f(x_1, \cdots, x_n) = 0$. Generalize the theory of tangents at simple points to this case. Show that the tangent hyperplane at a simple point $(c_1, \cdots, c_n)$ of the hypersurface is given by

$$\sum_{i=1}^{n} \left(\frac{\partial f}{\partial x_i}\right)_p (x_i - c_i) = 0.$$

$\left[\text{Hint: Use the hyperplane parametrization of §2. It}\right.$ will be necessary to use some other a_k in place of a_1 in that parametrization if $\left.\left(\frac{\partial f}{\partial x_1}\right)_p = 0.\right]$

7. Generalize the result in exercise 4 to hypersurfaces.
8. Curves of the form $y^2 = \prod_{i=1}^{n}(x - a_i)$ in $A_2(k)$ where the a_i are distinct are called *elliptic* if $n = 3, 4$ and *hyperelliptic* if $n > 4$. Take k to be the field of real numbers and choose $a_1 < \cdots < a_n$. Sketch these curves in the real euclidean plane for $n = 3, 4, 6$ and describe their behavior. In particular, derive a formula for the number of closed loops in the general case, for any n.
9. Suppose (x_0, y_0) is a double point of the curve $f(x, y) = 0$. Let $\Delta = \left(\frac{\partial^2 f}{\partial x \partial y}\right)^2 - \frac{\partial^2 f}{\partial x^2} \cdot \frac{\partial^2 f}{\partial y^2}$. Suppose k is the field of real numbers and that the curve is drawn in the real euclidean plane. Then (x_0, y_0) is a node, cusp, or isolated point according as $\Delta > 0$, $\Delta = 0$, or $\Delta < 0$ at (x_0, y_0). (It follows that the nodes of the curve are in the region of the plane in which $\Delta > 0$; corresponding statements hold for cusps and isolated points.)
10. Show that a curve in $A_2(k)$ cannot have an isolated point if k is algebraically closed.

II

PROJECTIVE SPACES

1. Definition of projective spaces

There are certain features of geometry in affine spaces that seem to carry us beyond the realm of affine spaces. In particular, we have in mind the notion of asymptote. Consider, for instance, the statement that the line $y = 0$ is an asymptote to the curve $xy = 1$ in the euclidean plane. This statement is usually explained in terms of the distance between a point on the curve and the x-axis approaching zero as we move along the curve in the (positive) x-direction. An asymptote may also be thought of as the limiting position of the tangent at a variable point p on the curve as p recedes indefinitely away from the origin. In any event the notion of limit obtrudes itself. The question arises whether it is possible to reformulate the theory of asymptotes in purely algebraic terms. This would be necessary to conform to our methodological requirements. Furthermore, it offers the only hope for generalization to arbitrary ground fields.

We have characterized an asymptote as the limiting position of a tangent. In other words, the asymptote is a sort of tangent to a point that "is not really there." What we do is throw some new points into our affine spaces so that they *are* there. When this is done the asymptotes behave like tangents at any other point; the algebraic formalism is exactly the same. This certainly would not be the case if we tried to do everything in purely affine terms; the asymptotes would be very special lines requiring a treatment different from that of tangents.

The fundamental idea underlying these new spaces can be phrased only vaguely now, but it will help us to understand what is going on. Later it will become clearer. Our first formulation of it is this: geometric loci in these spaces do not fly away indefinitely but remain within bounds so that the *whole* locus is "within reach," so to speak, all at once. There are "enough" points on the locus so that anything we wish to say about it can be said in terms of *points on* the locus.

Historically, there have been several motivations for the introduction of these so-called projective spaces. (As might be expected, one of them had to do with projections.) However, the sort of "completeness" concept we have just enunciated is the one that strikes closest to the heart of the matter in our present state of understanding. It is certainly the underlying idea in the use of these spaces in modern geometry.

We now have to decide how these spaces ought to be built. A point in $A_2(k)$ is given by a pair of numbers (x, y). Suppose we introduce new variables X, Y, Z and set $x = \frac{X}{Z}$, $y = \frac{Y}{Z}$. Then the triple (X, Y, Z) can be used to describe the point (x, y). Then X, Y, Z are called *homogeneous co-ordinates* of the point. So far, of course, we have to prohibit Z from being zero. Note also that if $t \neq 0$ in k then the triple (tX, tY, tZ) corresponds to the same point as (X, Y, Z); in other words, the homogeneous co-ordinates of a point are determined only to within a non-zero proportionality factor. Now suppose $f(x, y) = 0$ is the equation of an algebraic curve of degree n in $A_2(k)$. If we set $x = \frac{X}{Z}$ and $y = \frac{Y}{Z}$ and clear of denominators by multiplying through by Z^n, then we get a homogeneous equation (all terms of the same degree) in X, Y, Z. The equation of the curve in homogeneous co-ordinates becomes $F(X, Y, Z) = 0$ where $F(X, Y, Z) = Z^n \cdot f(X/Z, Y/Z)$ is a homogeneous polynomial of degree n.

So far we have restricted Z from being zero, since in this case a triple would not correspond to a point in $A_2(k)$. We now drop this restriction and admit triples (X, Y, Z) for which at least one of the components is not zero. If $Z = 0$, we call such a triple a *point at infinity*. These are the new points that we add to $A_2(k)$. The set of all triples (X, Y, Z), with not all components equal to zero, and proportional triples identified, we call *projective 2-space* $P_2(k)$ over k. We can think of $A_2(k)$ as being embedded in $P_2(k)$ as a subspace if we associate with a point (x, y) in $A_2(k)$ the point $(x, y, 1)$ in $P_2(k)$. In this case, the set of points (x, y, z) in $P_2(k)$ with $z = 0$ is called the *line at infinity*.

All this can be done in any number of dimensions. We define *projective n-space*, $P_n(k)$, as the set of $(n + 1)$-tuples $(x_0, x_1, \cdots, x_n)$ where the $x_i \in k$, not all components are zero, and proportional $(n + 1)$-tuples are identified. It is very important to notice that "points at infinity" are not distinguished in any way unless we are thinking of $A_n(k)$ as being embedded in $P_n(k)$ in some specific way, such as it was above in the case $n = 2$. The concept of a point being "at infinity" is meaningful only with regard to such a specific embedding. An *algebraic hypersurface* in $P_n(k)$ is the locus in $P_n(k)$ of a *homogeneous* polynomial equation $f(x_0, x_1, \cdots, x_n) = 0$ with coefficients in k. It is clear from this definition why we prohibit the $(n + 1)$-tuple with all components zero from being a point of $P_n(k)$; such a point would lie on *all* hypersurfaces—obviously an undesirable situation. About the requirement of homogeneity of the equation, there are two remarks to be made. First of all, if we start with a hypersurface in $A_n(k)$ and "homogenize" its equation as we did in the case $n = 2$, we end up with a homogeneous equation. Secondly, the co-ordinates of a point in $P_n(k)$ are determined only to within a proportionality factor and so it would be desirable for *all* the coordinates of a given point on the curve to satisfy the equation. It is easy to check (cf. Exercise 7) that this requires homogeneity of the equation provided the ground field k is "big enough."

2. Realizations of projective spaces

We wish now to consider two interpretations of projective space which are extremely helpful in studying certain aspects of the geometry of such spaces.

The first one is algebraic. The set of $(n + 1)$-tuples over a field k can be thought of as a vector space V_{n+1} of dimension $n + 1$ over k. The addition of two $(n + 1)$-tuples is defined by

$$(x_0, \cdots, x_n) + (y_0, \cdots, y_n) = (x_0 + y_0, \cdots, x_n + y_n)$$

and for $t \in k$ we define

$$t(x_0, \cdots, x_n) = (x_0, \cdots, x_n)t = (tx_0, \cdots, tx_n).$$

Points of $P_n(k)$ are given by $(n + 1)$-tuples with not all components zero and with proportional $(n + 1)$-tuples identified. In terms of the vector space structure, then, a point of $P_n(k)$ can be identified with a one-dimensional subspace of V_{n+1}. The importance of this realization of $P_n(k)$ lies in the fact that all the machinery of linear algebra becomes available to study these spaces. Actually, projective spaces are often *defined* in terms of the vector space interpretation. Lines are then defined as two-dimensional subspaces and hyperplanes as n-dimensional subspaces. Furthermore, the theory of linear transformation of V_{n+1} leads to the "appropriate" kinds of transformations to use on $P_n(k)$. This matter will be dealt with in §7.

Before leaving the subject, it is worthwhile to indicate how the vector space interpretation clarifies the sense in which projective spaces are "projective." We start with the set of bound geometric vectors in three-dimensional euclidean space. (As usual, "bound" means that all vectors emanate from the origin.) Suppose C is a curve in some plane P not containing the origin O. From this curve we generate a cone with apex at the origin. Now a point of $P_2(R)$ corresponds to a line through the origin. So, in particular, all points other than the origin on a generator of the cone represent the *same* point as a point of $P_2(R)$. Thus, if we cut this cone by another plane P' not through the origin, we get a curve of section C' which is really the *same* curve as a curve in $P_2(R)$.

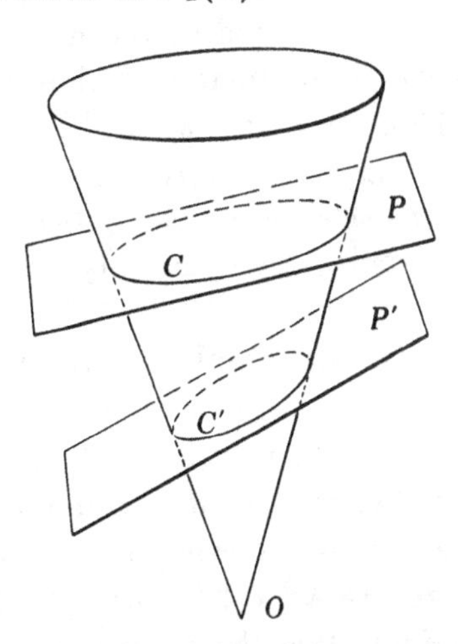

FIG. 2.1

If the curve C' is projected in the ordinary sense from O onto the plane P, the new curve C is identical with C' in the sense of $P_2(R)$. We have assumed here, since the purpose was only illustration, that C and C' are contained in some bounded region of three-dimensional euclidean space. (One also would suspect from these considerations that the "conics" in the projective plane are all pretty much the same. This suspicion is well founded (cf. §9).)

A second way of picturing a projective space is perhaps more geometric in the naïve sense. In contrast to the first, it concentrates more on the "boundedness" property of a projective space and suggests more clearly the way in which, as we said earlier, the whole curve comes within reach in such a space. In this case we work exclusively over the real numbers and confine attention to $P_2(R)$. The generalization to higher dimensions will be clear. Suppose (x, y, z) is a point in $P_2(R)$. Since $x^2 + y^2 + z^2 \neq 0$ we can multiply each component by a constant factor so that $x^2 + y^2 + z^2 = 1$. If this condition is satisfied, we shall say that the projective co-ordinates have been *normalized.* Even if this is done the normalized projective co-ordinates of a given point of $P_2(R)$ are still not uniquely determined. Suppose x, y, z are normalized co-ordinates of such a point. Any other normalized co-ordinates of this point are of the form tx, ty, tz. Then $x^2 + y^2 + z^2 = t^2(x^2 + y^2 + z^2) = 1$ and so $t = \pm 1$. The argument is reversible, and so for a given point of $P_2(R)$ there are exactly two ways of writing it in normalized co-ordinates — namely, (x, y, z) and $(-x, -y, -z)$.

We can now construct a model of $P_2(R)$ in three-dimensional euclidean space. Suppose p is a point in $P_2(R)$ given by (x, y, z) in normalized form. We can interpret x, y, z as co-ordinates of a point in three-dimensional euclidean space and thus p corresponds to a point $(x, y, z)_E$ on the sphere $X^2 + Y^2 + Z^2 = 1$. (We use the subscript E to indicate that the point is to be interpreted as a point of euclidean space.) Conversely, in this way, every point on the sphere determines exactly one point of $P_2(R)$. The correspondence fails to be one-to-one in that the points $(x, y, z)_E$ and $(-x, -y, -z)_E$ correspond to the same point of $P_2(R)$. Two such points on the sphere are distinguished by the fact that they are *antipodal*; that is, the line joining them goes through the center of the sphere.

Thus we have a one-to-one correspondence of $P_2(R)$ with the unit sphere in euclidean space where antipodal points are *identified*—that is, we count two such points as being the "same." This model is used most by topologists who usually have it in mind whenever they talk about projective spaces. The sort of identification used in this model is a standard procedure in topology. It is the same sort of idea we used in constructing the tori on which to represent finite affine spaces.

If anyone finds the procedure in the present case distasteful, it is quite easy to construct a "one-to-one" model simply by taking the upper hemisphere and half of the "equator."

3. Lines and hyperplanes

The equation of a line in affine 2-space becomes a homogeneous equation when written in terms of homogeneous co-ordinates, and similarly for higher dimensions. In view of this it is natural to define hyperplanes in $P_n(k)$ as the loci of homogeneous polynomial equations of degree 1. As usual we shall call these lines and planes in the cases $n = 2$ and $n = 3$ respectively. In §2 we gave another definition of hyperplane in terms of the vector space realization of $P_n(k)$. It is necessary, then, to show that the two definitions amount to the same thing. We shall see that the description of hyperplanes in terms of the vector space realization of $P_n(k)$ amounts to defining them by parametric equations.

Suppose $a_0x_0 + \cdots + a_nx_n = 0$ is the equation of a hyperplane in $P_n(k)$, according to the definition at the beginning of this section. Then some coefficient must be non-zero. Suppose $a_0 \neq 0$; the argument in other cases is similar. Then values of $x_1, \ldots, x_n$ may be assigned arbitrarily and then x_0 determined so that the equation is satisfied. Clearly we can determine n different $(n + 1)$-tuples

$v_j = (c_0^{(j)}, \cdots, c_n^{(j)})$ which satisfy the given equation and which are linearly independent as elements of V_{n+1}.

This means that in any relation $\sum_{j=1}^{n} t_j v_j = 0$ with $t_j \in k$,

all the t_j must be zero. Note that, in particular, no v_j can be the zero-vector and so must correspond to a point of $P_n(k)$. It is seen by direct substitution that any linear

combination $v = \sum_{j=1}^{n} t_j v_j$, $t_j \in k$, satisfies the equation

of the hyperplane; in other words, the subspace of V_{n+1} spanned by the v_j "lies in" the hyperplane. We contend that the hyperplane contains nothing else. To see this, suppose $(c_0, \cdots, c_n)$ is any point on the hyperplane. If this, as an element of V_{n+1}, is not linearly dependent on the v_j, then it together with the v_j must span the whole space. (We have used here the fact that any $n + 1$ linearly independent vectors of V_{n+1} must span the whole space.) However, this state of affairs is untenable since we can easily construct a point of $P_n(k)$ not the hyperplane. Indeed, if $a_0 \neq 0$ then the point $(1, 0, \cdots, 0)$ will not be on the hyperplane. It follows that we can find n points $(c_0^{(j)}, \cdots, c_n^{(j)})$ on the hyperplane, $j = 1, \cdots, n$, which are linearly independent as vectors in V_{n+1} and having the following property: a point $(x_0, \cdots, x_n)$ is on the hyperplane if and only if

$$(x_0, \cdots, x_n) = \sum_{j=1}^{n} t_j (c_0^{(j)}, \cdots, c_n^{(j)});$$

that is, $x_i = \sum_{j=1}^{n} t_j c_i^{(j)}$ for $i = 0, 1, \cdots, n$ and $t_j \in k$.

In the case $n = 2$, the linear independence of v_1 and v_2 is equivalent to the statement that the corresponding points in $P_2(k)$ are distinct.

Now suppose we are given n linearly independent $(n+1)$-tuples $v_j = (c_0^{(j)}, \cdots, c_n^{(j)})$, $j = 1, 2, \cdots, n$. We want to show that the points of $P_n(k)$ given by the parametric equation $(x_0, \cdots, x_n) = \sum_{j=1}^{n} t_j(c_0^{(j)}, \cdots, c_n^{(j)})$ constitute precisely the locus of a homogeneous polynomial equation of degree 1. Consider the equation $a_0x_0 + \cdots + a_nx_n = 0$ where the coefficients a_i are to be determined. If we substitute in turn each of the vectors $v_1, \cdots, v_n$ we obtain n linear equations in $n+1$ unknowns $a_0, \cdots, a_n$. Such a system always has a non-trivial solution and so $v_1, \cdots, v_n$ correspond to points on a certain hyperplane. The argument we used before shows that the only points on the hyperplane obtained in this way correspond to linear combinations of the v_j. This completes the proof that the two definitions of hyperplane are equivalent.

Remark: By similar methods we can obtain a "decent" parametrization of hyperplanes in affine spaces. (What would this be?) The parametrization in Chapter I was used since its derivation required nothing from linear algebra. In projective spaces this issue of linear algebra can no longer be evaded. We have used only the elementary part of the theory relevant to the notion of linear dependence which is much simpler than introducing determinants.

4. Intersections of lines with algebraic curves

Our purpose in this section is to generalize our results on intersections of lines and curves in affine spaces to the projective case. The main tool will again be the formal Maclaurin expansion. In the projective case we deal in particular with homogeneous polynomials and shall have occasion to make frequent use of Euler's Theorem on homogeneous functions. Here, of course, we shall use it only in the case of homogeneous polynomials when the derivatives are to be interpreted in a purely algebraic way, as in the previous chapter. The statement of Euler's Theorem is this: if $f(x_1, \cdots, x_m)$ is a homogeneous polynomial of degree n then $nf = \sum_{i=1}^{m} \frac{\partial f}{\partial x_i} \cdot x_i$. It is proved by direct computation (cf. Exercise 4).

Now let $f(x_1, x_2, x_3) = 0$ be the equation of an algebraic curve C in $P_2(k)$. It will be assumed that the degree n of $f(x_1, x_2, x_3)$ is less than the characteristic of k in case k has a non-zero characteristic. (The case where the characteristic of k is less than n can be treated by a method similar to that used in the affine case.)

Suppose $p = (a_1, a_2, a_3)$ is a point on C. We wish to study the way in which lines through p intersect C. Such a line is determined by giving a point (b_1, b_2, b_3) on it which is distinct from p, and then all points on the line are given by $(sa_1 + tb_1, sa_2 + tb_2, sa_3 + tb_3)$ where s and t are in k. Call this line L. The intersections of L and C are obtained by finding the values of s and t such that

$$(1) \qquad f(sa_1 + tb_1, sa_2 + tb_2, sa_3 + tb_3) = 0.$$

The intersection point p corresponds to the case $s = 1$, $t = 0$. Let us consider the left-hand side of (1) as a function of t alone and denote it by $F(t)$. We now expand $F(t)$ in its formal Maclaurin expansion

$$(2) \quad F(t) = F(0) + \frac{F'(0)}{1!}t + \frac{F''(0)}{2!}t^2 + \cdots + \frac{F^{(n)}(0)}{n!}t^n.$$

As in the affine case, we say that L *intersects* C *at* p *with multiplicity* m if $F^{(m)}(0) \neq 0$ and $F^{(l)}(0) = 0$ for $l < m$. It is easy to show, as in the affine case, that this definition depends on the line L and not on any particular parametrization of L.

Now suppose L intersects C at p with multiplicity $m \geq 2$. This means $F'(0) = 0$. Going back to (1), we see that this is synonymous with the condition $\sum_{i=1}^{3}\left(\frac{\partial f}{\partial x_i}\right)_p b_i = 0$. As usual, the subscript on the derivative means that it is to be evaluated at p. If $\left(\frac{\partial f}{\partial x_i}\right)_p = 0$ for $i = 1,2,3$ then p is called a *singular* point; otherwise it is a *simple* point. Suppose p is simple. Then the condition means that the point (b_1, b_2, b_3) must lie on the line $\sum_{i=1}^{3}\left(\frac{\partial f}{\partial x_i}\right)_p x_i = 0$. By using Euler's Theorem it is verified immediately that p lies on this line. Thus we see that if p is a simple point, there is a unique line cutting C at p with multiplicity $m \geq 2$ and its equation is

$$\sum_{i=1}^{3}\left(\frac{\partial f}{\partial x_i}\right)_p x_i = 0. \tag{3}$$

We call this line the *tangent* at p. Indeed, we define tangents in terms of multiplicities for the projective case in exactly the same way as we did for the affine case.

At this point the natural question arises as to whether what we have done so far is consistent with the results obtained for the affine case. It is. To see this, let us start with a curve $f(x, y) = 0$ of degree n in $A_2(k)$ and suppose $p = (x_0, y_0)$ is a point on the curve. We homogenize $f(x, y)$ as usual by setting $F(x_1, x_2, x_3) = x_3^n f\left(\frac{x_1}{x_3}, \frac{x_2}{x_3}\right)$ and we write the homogeneous co-ordinates of p in the *particular* form $(x_0, y_0, 1)$. Then one checks immediately that $\left(\frac{\partial F}{\partial x_1}\right)_{(x_0,y_0,1)} = \left(\frac{\partial f}{\partial x}\right)_p$ and $\left(\frac{\partial F}{\partial x_2}\right)_{(x_0,y_0,1)} = \left(\frac{\partial f}{\partial y}\right)_p$. Also, applying Euler's Theorem to $F(x_1, x_2, x_3)$, we see that

$$\left(\frac{\partial F}{\partial x_3}\right)_{(x_0,y_0,1)} \cdot 1 = -\left(\frac{\partial F}{\partial x_1}\right)_{(x_0,y_0,1)} \cdot x_0 - \left(\frac{\partial F}{\partial x_2}\right)_{(x_0,y_0,1)} \cdot y_0.$$

Using these results, we see that p is singular in the affine sense if and only if it is singular in the projective sense; also that if p is simple, the equation (3) reduces to the equation for the tangent in the affine case.

All our considerations so far can be generalized to hypersurfaces in $P_n(k)$ if we use the parametrization of hyperplanes introduced in §3. If $f(x_0, \cdots, x_n) = 0$ is an algebraic hypersurface in $P_n(k)$ and $p = (a_0, \cdots, a_n)$ is a point on it, we say that p is *singular* if $\left(\frac{\partial f}{\partial x_i}\right)_p = 0$ for $i = 0, 1, \cdots, n$; otherwise it is *simple*. If p is simple there is a unique tangent hyperplane given by the equation $\sum_{i=0}^{n} \left(\frac{\partial f}{\partial x_i}\right)_p x_i = 0.$

To go back to curves again, we can discuss singular points in a way similar to that used in the affine case. We say p is an *r-fold singular point* of C if all partials of order strictly less than r vanish at p, but some partial of order r does not. We shall discuss only the case $r = 2$ to show how to cope with the slight difference in formalism. Suppose that p is a double point, that is, that $\left(\frac{\partial f}{\partial x_i}\right)_p = 0$ for $i =$ 1, 2, 3 and that some second-order partial does not vanish at p. Then any line through p cuts C there with multiplicity at least 2. The condition for this multiplicity to be at least 3 is $F''(0) = 0$. This means that

$$\sum_{i,j=1}^{3} \left(\frac{\partial^2 f}{\partial x_i \partial x_j}\right)_p b_i b_j = 0.$$

Thus the point (b_1, b_2, b_3) must lie on the locus:

$$\sum_{i,j=1}^{3}\left(\frac{\partial^2 f}{\partial x_i \partial x_j}\right)_p x_i x_j = 0. \tag{4}$$

We contend that the locus (4) consists of either the point p alone, or straight lines through p. To see this, apply Euler's Theorem to the polynomials $\dfrac{\partial f}{\partial x_i}$ $(i = 1, 2, 3)$. This gives

$$(n-1)\frac{\partial f}{\partial x_i} = \sum_{j=1}^{3} \frac{\partial^2 f}{\partial x_i \partial x_j} \cdot x_j. \tag{5}$$

Substituting $x_i = a_i$ $(i = 1, 2, 3)$ we obtain

$$(n-1)\left(\frac{\partial f}{\partial x_i}\right)_p = \sum_{j=1}^{3}\left(\frac{\partial^2 f}{\partial x_i \partial x_j}\right)_p a_j.$$

Thus

$$\sum_{i,j=1}^{3}\left(\frac{\partial^2 f}{\partial x_i \partial x_j}\right)_p a_i a_j = (n-1)\sum_{i=1}^{3}\left(\frac{\partial f}{\partial x_i}\right)_p a_i = 0$$

since p is a singular point. Therefore p is on the locus (4). If (c_1, c_2, c_3) is any other point on (4), then a simple computation using (5) shows that any point of the form $(sa_1 + tc_1, sa_2 + tc_2, sa_3 + tc_3)$ is also on (4). Thus the locus (4) consists of straight lines if it contains any point other than p itself. The point p is called a *cusp, node,* or *isolated point* according as (4) is the equation of one line, two lines, or the single point p. In the first two cases, equation (4) gives the tangent(s) at p. (That the locus (4) consists of two straight lines *at most,* follows from §6.)

5. Asymptotes

Suppose C is a curve in $A_2(k)$ given by a polynomial equation $f(x, y) = 0$ of degree n. This equation is "homogenized" by setting $x = \frac{x_1}{x_3}$ and $y = \frac{x_2}{x_3}$. Then the polynomial $F(x_1, x_2, x_3) = x_3^n f\left(\frac{x_1}{x_3}, \frac{x_2}{x_3}\right)$ is homogeneous of degree n. The curve given by $F(x_1, x_2, x_3) = 0$ in $P_2(k)$ is called the projective curve C' corresponding to C. The curve C' may or may not intersect the line at infinity, $x_3 = 0$. Suppose it does and suppose that p is a point at infinity on C'. We can then construct the tangent (or tangents) at p provided it is not an isolated point. Such a tangent may lie entirely in, and so coincide with, the line at infinity. Otherwise, the points of this line which lie in $A_2(k)$ constitute a line in $A_2(k)$. Such a line in $A_2(k)$ is called an *asymptote* to the curve. In other words, an asymptote to a curve in $A_2(k)$ is a line which is tangent to the corresponding projective curve at a point at infinity on the latter curve.

As an example, consider the curve $xy - 1 = 0$ in $A_2(k)$ where k is the field of real numbers. The corresponding projective curve has the equation $F(x_1, x_2, x_3) \equiv x_1 x_2 - x_3^2 = 0$. The intersection points with the line at infinity are obtained by setting $x_3 = 0$ and solving for x_1 and x_2. We get two such points, (1, 0, 0) and (0, 1, 0). Let us call p the first of these points. It is a simple point, and equation (3) gives $x_2 = 0$ as the equation of the tangent at p. The affine part of this line is given by $y = 0$, and so this is an asymptote. Similarly, the tangent at (0, 1, 0) gives the asymptote $x = 0$.

It is important to realize that the concept of asymptote is meaningful only in terms of the embedding of some specific $A_2(k)$ in $P_2(k)$ so that the points at infinity are distinguished. In a purely projective framework, the term "asymptote" is meaningless.

The question of pictorial representation arises again in connection with asymptotes, since we have defined them in a purely algebraic way. This question arises, of course, only in the case of affine curves over the field of real numbers when they can be interpreted as curves in the euclidean plane. As in the comparable discussion of tangents in Chapter I, we shall not attempt a really precise argument, but rather only indicate the main idea sufficiently to attain plausibility and to make it clear how to construct a rigorous proof. Suppose we have a curve in the euclidean plane with a branch ("part") that extends away indefinitely from the origin. Suppose the tangent to a variable point p on this branch approaches a limiting position as p moves away indefinitely from the origin. This limiting line is what we usually mean by an asymptote in its intuitive sense.

Let x_1, x_2, x_3 be homogeneous co-ordinates of the variable point p. We saw in §2 that these can be normalized by the requirement that $x_1^2 + x_2^2 + x_3^2 = 1$. If p is a point in the *affine* part of the plane, then $x_3 \neq 0$ and, since we are working over the real numbers, we can impose the further requirement that x_3 be *positive*. If this is done, we shall say that the homogeneous co-ordinates are *strongly normalized*. This removes the "twofold ambiguity" of the normalized co-ordinates for points not at infinity; the crucial fact is that the strongly normalized co-ordinates of the point in $A_2(R)$ are *unique*. Now as p moves along the branch away from the origin we must have $\dfrac{x_1^2 + x_2^2}{x_3^2} \longrightarrow +\infty$. Thus $x_3 \longrightarrow 0$ and x_1 and x_2 must approach definite values, not both zero. Thus p approaches a point at infinity p_0 on the corresponding curve. The coefficients of the tangent equation at p depend continuously on the strongly normalized co-ordinates of p and so the tangent at p in the algebraic sense approaches as limit the tangent at p_0, which gives the asymptote to the branch in question.

Remark: The existence of the limit point p_0 in the argument above is one of the most important properties of real projective spaces. It is a consequence of the topological statement that the n-sphere is compact. The connection lies in the representation of a real projective space as such a sphere with its antipodal points identified.

Example 1: Let C be the curve $y = \dfrac{1}{1 + x^2}$ in $A_2(k)$ where k is the field of real numbers (Fig. 2-2). The corresponding projective curve has the equation $f(x_1, x_2, x_3) = 0$ where $f(x_1, x_2, x_3) \equiv x_2x_3^2 + x_2x_1^2 - x_3^3$. To find the points of infinity on the curve we intersect it with the line at infinity $x_3 = 0$. We obtain two points, $(1, 0, 0)$ and $(0, 1, 0)$. The first of these is a simple point and the tangent there is $x_2 = 0$. The affine part of this is the line $y = 0$ in $A_2(k)$ and this line is an asymptote to the affine curve. The point $(0, 1, 0)$ is a singular point since all the first partials vanish there. The only second partials which do not vanish there are $\dfrac{\partial^2 f}{\partial x_1{}^2}$ and $\dfrac{\partial^2 f}{\partial x_3{}^2}$, both of which take on the value 2. Equation (4) reduces to $x_1{}^2 + x_3{}^2 = 0$. The locus of this equation consists only of the single point $(0, 1, 0)$ and so this is an isolated point, a curious state of affairs in view of the curve's "exemplary" behavior in the affine plane.

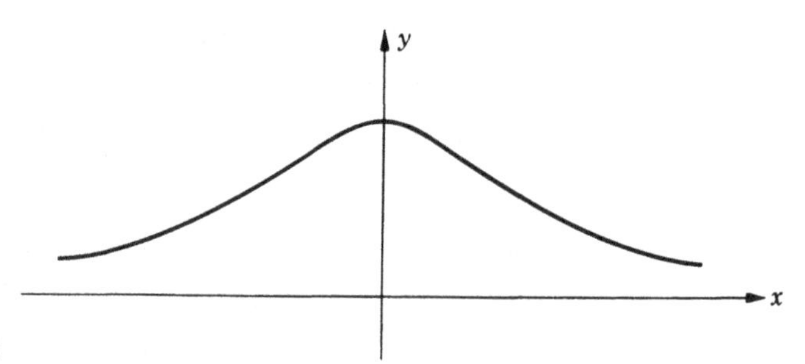

FIG. 2.2

Example 2: It would be interesting to see how it is that curves that "should not" have asymptotes indeed do not.

A very simple case is the curve $x^2 + y^2 = 1$ in the real affine plane. In homogeneous form the equation becomes $x_1^2 + x_2^2 = x_3^2$. If we set $x_3 = 0$ then so are x_1 and x_2. Therefore the curve does not intersect the line at infinity, and so lies entirely in the affine plane. Thus there is no question of tangents at points at infinity.

Example 3: Another more interesting case is the curve $y = x^3$ in the real affine plane. The corresponding projective curve has the equation $f(x_1, x_2, x_3) = 0$ where $f(x_1, x_2, x_3) \equiv x_1^3 - x_2 x_3^2$. Then $\frac{\partial f}{\partial x_1} = 3x_1^2$, $\frac{\partial f}{\partial x_2} = -x_3^2$, and $\frac{\partial f}{\partial x_3} = -2x_2 x_3$. The only intersection with the line at infinity is the point $(0, 1, 0)$. This is a singular point, indeed the only one on the whole curve. The only second partial that does not vanish at this point is $\frac{\partial^2 f}{\partial x_3^2}$. Thus $(0, 1, 0)$ is a cusp and the tangent there is given by $x_3^2 = 0$, that is, $x_3 = 0$. This line contains no points of the affine plane, and for this reason the affine curve has no asymptote.

6. Further remarks on intersections of lines and curves

We collect here some miscellaneous results that will be used later. The problem with which we concern ourselves is the determination of the number of intersections of a line with a curve of degree n. In particular we wish to determine conditions under which it is possible to say that there will be n such intersections.

Proposition 1: Suppose k is an infinite field and that $f(x_1, x_2, x_3) = 0$ is a curve of degree n in $P_2(k)$. Assume that this curve contains the line

$$p(x_1, x_2, x_3) \equiv \sum_{i=1}^{3} a_i x_i = 0.$$

Then there is a polynomial $g(x_1, x_2, x_3)$ such that $f(x_1, x_2, x_3) = p(x_1, x_2, x_3) \cdot g(x_1, x_2, x_3)$.
Proof: Suppose $a_1 \neq 0$. (The proof in the other cases is similar.) Define new variables $X_1 = p(x_1, x_2, x_3)$, $X_2 = x_2$, and $X_3 = x_3$. This transformation is clearly invertible. The polynomial $f(x_1, x_2, x_3)$ becomes a polynomial $F(X_1, X_2, X_3)$ and the hypothesis implies that $F(0, X_2, X_3)$ is identically zero as a polynomial function. Since k is infinite, this implies that X_1 divides $F(X_1, X_2, X_3)$. The proposition follows directly upon interpreting this statement in terms of the original variables.

Proposition 2: Suppose that k is algebraically closed. Let C be a curve and L a line in $P_2(k)$. Then L and C intersect.
Proof: Suppose C has the equation $f(x_1, x_2, x_3) = 0$ and L is given in parametric form $x_i = sa_i + tb_i$, $i = 1, 2, 3$ as in §4. The intersections are given by the solutions in s and t to the equation

$$f(sa_1 + tb_1, sa_2 + tb_2, sa_3 + tb_3) \equiv F(s, t) = 0.$$

If $s = 0$ satisfies this equation, then (b_1, b_2, b_3) is an intersection point. Otherwise set $s = 1$ and solve for t. A solution exists by virtue of the algebraic closure of k. This completes the proof.

We now investigate these intersections more thoroughly, particularly with a view to their "enumeration." Suppose that C has degree n and that L does not lie in C. If (a_1, a_2, a_3) is an intersection point and (b_1, b_2, b_3) is a point of L not on C, then all the intersections are obtained by solving the equation $F(1, t) = 0$ since the case $s = 0$ does not give an intersection. Since (b_1, b_2, b_3) is not on

C it follows, from examining the terms of highest degree, that the equation $F(1, t) = 0$ has degree n. Since k is algebraically closed, there are exactly n roots, provided multiplicities are counted as in the theory of equations. Note that the multiplicity of the root $t = 0$ coincides with the intersection multiplicity of L with C at (a_1, a_2, a_3) as defined earlier. (This is true even when the characteristic of k is less than or equal to n (cf. §5 of Chapter I).) If the intersections of L and C are counted with the multiplicities of the corresponding roots of $F(1, t) = 0$, there are exactly n of them. We have proved

Proposition 3: Let C be a curve of degree n and L a line in $P_2(k)$. If L does not lie in C then there are at most n intersection points. If k is algebraically closed, there are exactly n intersections when multiplicities are counted as described above.

Remarks: This result is very different from what happens in $A_2(k)$, in which case a curve and a line may not intersect; for example, the two lines $x = 1$ and $x = 2$. Even when they do, there may not be "enough" intersection points. For instance, the line $x = 0$ cuts the curve $y = x^3$ only at the origin and with multiplicity 1. However, for the corresponding projective line $x_1 = 0$ and curve $x_1^3 - x_2x_3^2 = 0$ in $P_2(k)$ we get the intersections (0, 0, 1) with multiplicity 1 and (0, 1, 0) with multiplicity 2.

Proposition 4: Suppose C is a curve of degree $n > 1$ in $P_2(k)$ where k is algebraically closed. Suppose C contains a line L. Then C has singular points.

Proof: We can write

$$f(x_1, x_2, x_3) = p(x_1, x_2, x_3) \cdot g(x_1, x_2, x_3)$$

as in Proposition 1. Then $\dfrac{\partial f}{\partial x_i} = p\dfrac{\partial g}{\partial x_i} + g\dfrac{\partial p}{\partial x_i}$ for $i = 1, 2, 3$. Let (c_1, c_2, c_3) be a point on the intersection of $p(x_1, x_2, x_3) = 0$ and $g(x_1, x_2, x_3) = 0$. Such a point exists by Proposition 2. Obviously $\frac{\partial f}{\partial x_i} = 0$ at this point for $i = 1,2,3$ and so it is a singular point of C.

7. Projective transformations

Suppose $P_n(k)$ is realized as the set of one-dimensional subspaces of the vector space V_{n+1} of $(n + 1)$-tuples over k. An *invertible* linear transformation ϕ of V_{n+1} induces a transformation of $P_n(k)$ into itself since if $(x_0, \cdots, x_n)$ is a vector in V_{n+1} with some $x_i \neq 0$ then $\phi(x_0, \cdots, x_n) = (y_0, \cdots, y_n)$ with some $y_j \neq 0$. Thus if $(x_0, \cdots, x_n)$ is interpreted as a point in $P_n(k)$, then also $(y_0, \cdots, y_n)$ will be a point in $P_n(k)$. If $\lambda \neq 0$ in k, then ϕ and $\lambda\phi$ induce the same transformation of $P_n(k)$. If we denote this induced transformation of $P_n(k)$ by Φ, then Φ maps $(x_0, \cdots, x_n)$ into $(y_0, \cdots, y_n)$ where $\rho y_j = \sum_{i=0}^{n} x_i a_{ij}$ with some $\rho \neq 0$ in k and $\det(a_{ij}) \neq 0$. Transformations Φ of this type are called *projective transformations*. It often is convenient to write the preceding equations in matrix form:

$$\rho(y_0, \cdots, y_n) = (x_0, \cdots, x_n)\begin{pmatrix} a_{00} & \cdots & a_{0n} \\ \cdot & & \cdot \\ \cdot & & \cdot \\ \cdot & & \cdot \\ a_{n0} & \cdots & a_{nn} \end{pmatrix}.$$

A projective transformation Φ of $P_n(k)$ preserves incidence properties in the sense that points are mapped into points, lines into lines, planes into planes, etc., and intersection properties are preserved. For example, if L_1 and L_2 are two lines of $P_n(k)$ intersecting in a point p then $\Phi(L_1)$ and $\Phi(L_2)$ intersect in $\Phi(p)$. All these properties are just translations into "projective" language of the fact that the corresponding linear transformation ϕ of V_{n+1} maps r-dimensional subspaces into r-dimensional subspaces, and their intersections into the intersections of the images. Since ϕ is an invertible linear transformation it follows that projective transformations are also invertible. Indeed the projective transformations of $P_n(k)$ form a group under multiplication. (If Φ and Ψ are projective transformations, the product $\Phi \circ \Psi$ is defined by $\Phi \circ \Psi : p \longrightarrow \Phi(\Psi(p))$.)

(For readers familiar with the notion of quotient group, the group of projective transformations of $P_n(k)$ can easily be described in matrix language: let G be the group of invertible matrices over k of degree $n + 1$ and let H be the subgroup consisting of matrices of the form $\lambda \cdot I$ where $\lambda \neq 0$ in k and I is the identity matrix of degree $n + 1$. Then H is a normal subgroup of G and the group of projective transformations is isomorphic to the quotient group $\frac{G}{H}$.)

Remark: The equations describing projective transformations are obviously generalized forms of similar equations used in elementary analytic geometry to "change axes" by rotation or translation. From the point of view adopted in this book, transformations are regarded as moving the *points* around rather than changing axes; this, primarily, for the reason that we *have* no axes. We have not set up points as kinds of "ideal" objects independent of their representation in a given co-ordinate system, but rather merely as n-tuples from a field.

8. Quadrics

A quadric is a hypersurface in $P_n(k)$ given by a polynomial equation of the second degree. In the case $n = 2$, it is called a conic and in the case $n = 3$, a quadric surface. In higher dimensions, quadrics should probably be called hyperquadrics, but it is customary among geometers to omit the prefix, since its only function would seem to consist in evoking the respectful humility of the uninitiated. For technical reasons, *it will be assumed throughout this section that the characteristic of the field is not 2.* (Recall that this means $1 + 1 \neq 0$ in k.) With this assumption, every quadric V in $P_n(k)$ can be given by an equation of the form

$$(1) \qquad f(x_0, \cdots, x_n) \equiv \sum_{i,j=0}^{n} a_{ij} x_i x_j = 0,$$

$a_{ij} = a_{ji} \in k$ (cf. exercise 8). Computing the partials, we find that $\dfrac{\partial f}{\partial x_i} = 2 \sum_{j=1}^{n} a_{ij} x_j$. Thus V has a singular point if and only if the set of equations

$$\sum_{j=1}^{n} a_{ij} x_j, \; i = 0, \cdots, n$$

has a non-trivial solution. This proves

Theorem 1: The quadric (1), assuming the characteristic of k is not 2, has a singular point if and only if $\det(a_{ij}) = 0$.

Now suppose $(c_0, \cdots, c_n)$ is a singular point of V and that $(d_0, \cdots, d_n)$ is any other point of V. Then the line joining these two points lies entirely in V. To see this, let λ, μ be any two elements of k. Then

$$\sum_{i,j=0}^{n} a_{ij}(\lambda c_i + \mu d_i)(\lambda c_j + \mu d_j)$$

$$= \sum_{i=0}^{n} (\lambda c_i + \mu d_i) \sum_{j=0}^{n} a_{ij}(\lambda c_j + \mu d_j)$$

$$= \sum_{i=0}^{n} (\lambda c_i + \mu d_i) \sum_{j=0}^{n} a_{ij} \cdot \mu d_j \left(\text{since } \sum_{j=0}^{n} a_{ij} c_j = 0\right)$$

$$= \sum_{i,j=0}^{n} \lambda \mu a_{ij} c_i d_j + \mu^2 \sum_{i,j=0}^{n} a_{ij} d_i d_j.$$

The first term here is equal to $\lambda\mu \sum_{i,j=0}^{n} a_{ij}c_jd_i$ since the

a_{ij} are symmetric in their indices, and $\sum_{j=0}^{n} a_{ij}c_j = 0$ for

each i. The second term is zero since $(d_0, \cdots, d_n)$ is on V.

This result has important consequences for the case of conics. Suppose V is a conic in $P_2(k)$ and that k is algebraically closed. Then V contains at least two points. It follows that if V contains a singular point then it consists of two (possibly coincident) straight lines. The converse follows by Proposition 4 of §6. These facts can be summarized by the following:

Theorem 2: Suppose C is a conic in $P_2(k)$ where k is an algebraically closed field of characteristic not 2. Then the following statements are equivalent:

(i) C has a singular point.
(ii) Det $(a_{ij}) = 0$.
(iii) C consists of two (possibly coincident) straight lines.

9. Canonical form of quadrics in projective spaces

Let V be the quadric in $P_n(k)$ given by the equation

$$f(x_0, \cdots, x_n) \equiv \sum_{i,j=0}^{n} a_{ij}x_ix_j = 0 \text{ with } a_{ij} = a_{ji} \in k.$$

Assume as usual that k does not have characteristic 2. We will show that there exists a projective transformation of $P_n(k)$ which maps this quadric into one of the form

$$\sum_{i=0}^{n} a_i x_i^2 = 0, \text{ where } a_i \in k.$$

The quadric is then said to be in *canonical form*. (This means that this form is a restricted type of equation to which all others of the class may be reduced.) The reduction to this form is usually accomplished by translating the problem into matrix theory and using certain elementary results on the diagonalization of symmetric matrices. As in some other matters, however, we shall go at it in a very naïve way, as a sort of glorious exercise in completing squares.

First of all, we reduce the problem to the case $a_{00} \neq 0$:

(i) If $a_{00} = 0$ and $a_{kk} \neq 0$ for some k, then we set $X_0 = x_k$, $X_k = x_0$, and $X_i = x_i$ for $i \neq 0, k$. This, of course, amounts to a projective transformation.

(ii) If all $a_{ii} = 0$ then some $a_{kl} \neq 0$ with $k \neq l$ and by a transformation of the previous type, we are reduced to the case $a_{01} \neq 0$. Then set $X_0 = x_0$, $X_1 = x_1 - x_0$, and $X_i = x_i$ for $i \neq 0,1$. Then $\sum_{i,j=0}^{n} a_{ij}x_ix_j$ transforms into $\sum_{i,j=1}^{n} a'_{ij}X_iX_j$ where $a'_{00} = 2a_{01}$. It is assumed that k does not have characteristic 2, and so $a'_{00} \neq 0$. Now we reduced to the case where in the original equation $a_{00} \neq 0$.

(iii) The next step consists of "completing the square" in x_0. We start from the observation that

$$a_{00}^{-1}\left[a_{00}x_0 + \sum_{i=1}^{n} a_{i0}x_i\right]^2$$

$$= a_{00}x_0^2 + 2\sum_{i=1}^{n} a_{i0}x_0x_i + \sum_{i,j=1}^{n} a_{00}^{-1}a_{i0}a_{0j}x_ix_j.$$

Now set

$$\begin{cases} X_0 = x_0 + a_{00}^{-1}\sum_{i=1}^{n} a_{i0}x_i \\ X_i = x_i, \quad \text{for} \quad i \geq 1. \end{cases}$$

Making this substitution in the polynomial $f(x_0, \cdots, x_n)$ we obtain $a_{00}X_0^2 + g(X_1, \cdots, X_n)$ where $g(X_1, \cdots, X_n)$ is of the form

$$\sum_{i,j=1}^{n} b_{ij} X_i X_j \text{ with } b_{ij} = b_{ji} \in k.$$

(iv) We now repeat the whole process from the beginning for $g(X_1, \cdots, X_n)$ and continue in this way until the reduction is complete.

The quadric now has the desired form

$$\sum_{i=0}^{n} a_i X_i^2 = 0.$$

The determinant of coefficients is just $\prod_{i=0}^{n} a_i$ and the quadric has a singular point if and only if this product is zero.

Observe that if k is algebraically closed then each coefficient a_i has a square root and so a further transformation of the form

$$X_i = a_i^{-\frac{1}{2}} x_i \text{ for } a_i \neq 0$$

$$X_i = x_i \qquad \text{for } a_i = 0$$

changes the equation to the form

$$\sum x_i^2 = 0,$$

where the sum is over some of the integers $0, 1, 2, \cdots, n$. Clearly it is possible, by a very trivial transformation, to make the summation range consist of $0, 1, \cdots, r$ where $r \leq n$. In particular if k is algebraically closed and not of characteristic 2, there is—to within a projective transformation—exactly one non-singular quadric in $P_n(k)$ and its equation is $\sum_{i=0}^{n} x_i^2 = 0$.

Exercises for Chapter II

1. Find the number of points in $P_n(k)$ where k is the field F_p of residue-classes of integers modulo a prime p.
2. Analyze as far as possible the following curves in $A_2(k)$ when k is the field of real numbers. Find any singular points and determine their nature (for the whole corresponding projective curve) and sketch them.

   (i) Strophoid $y^2 = x^2\left(\frac{a + x}{a - x}\right)$ $(a > 0)$.

   (ii) Cissoid $y^2 = \frac{x^3}{a - x}$ $(a > 0)$.

   (iii) Limaçon $\rho = b - a \cos \theta$ $(0 < b < a)$.

   (iv) Lemniscate $\rho^2 = a^2 \cos 2\theta$ $(a > 0)$.

   (v) $y^2 = \frac{x^2}{(x + 1)(x + 2)}$.
3. Show that the elliptic curve $y^2 = (x - a_1)(x - a_2)(x - a_3)$ in $A_2(C)$, where a_1, a_2, a_3 are distinct complex numbers, has no singularities, even for the corresponding curve in $P_2(C)$.
4. Prove Euler's Theorem: if f is a homogeneous polynomial of degree n in variables $x_1, \cdots, x_m$ then
   $$nf = \sum_{i=1}^{m} \frac{\partial f}{\partial x_i} x_i.$$
5. Show that intersection multiplicities for lines and curves, and the property of tangency, are preserved by projective transformations of $P_2(k)$.
6. Show that if $f(x_1, x_2, x_3)$ is a polynomial over an algebraically closed field k, then there exists a point in $P_2(k)$ such that some representation of it as an $(n + 1)$-tuple satisfies the equation $f(x_1, x_2, x_3) = 0$. Generalize to $P_n(k)$.
7. Same assumptions as in exercise 6. Show that if for every point p on the locus, *every* set of (homogeneous)

co-ordinates of p satisfied the equation, then the equation must be homogeneous.

8. Explain why a quadric in $P_n(k)$ can always be given an equation of the form $\sum_{i,j=0}^{n} a_{ij}x_ix_j = 0$ with $a_{ij} = a_{ji} \in k$ if k has a characteristic other than 2. Explain why this is not so if the characteristic of k *is* 2.

III

RATIONAL CURVES

In elementary geometry of the euclidean plane, it often happens that a curve is given by parametric equations. Such a parametrization is frequently a help in studying the geometry of the curve. In the case of algebraic curves, of course, we wish to restrict the class of functions used in the parametrization so that purely algebraic methods can be employed. Suppose C is an algebraic curve in $A_2(k)$. Then C is called a *rational curve* if it can be parametrized in the form $x = f(t)$, $y = g(t)$ where $f(t)$ and $g(t)$ are rational functions of a variable t. (Recall that by "rational function" is meant a quotient of polynomial functions.) We shall allow a finite number of exceptions in the sense that at most a finite number of points of the curve are not obtained by taking a suitable value of t in k. Having allowed this, it would be rather silly to work over finite fields, so *it will be assumed that k is infinite*. Rational curves are sometimes called *unicursal curves*, especially in the older literature. (The reason for the choice of this word should be clear by the end of this chapter, to anyone with a modest knowledge of Latin.) The notion of rationality carries over to any number of dimensions; a hypersurface in $A_n(k)$ is said to be rational if the co-ordinates of its points are given by rational functions of $n - 1$ independent variables ranging over k, again with certain "reasonable" exceptions (cf. Chapter V).

We shall restrict ourselves entirely to the case of curves and give some attention to how the exceptions arise in a rational parametrization. This in turn forces us to extend our considerations to the projective case and to algebraically closed fields.

The subject of rational curves also leads in a natural way to some aspects of the connection of algebraic geometry with number theory. The particular topic we shall discuss from this point of view is that of diophantine equations. Suppose $f(x_1, \cdots, x_n)$ is a polynomial with rational integral coefficients. The problem to be solved is that of finding all solutions of the equation $f(x_1, \cdots, x_n) = 0$ in which the x_i are rational integers. An equation of this form for which we wish to obtain integral solutions is call called a *diophantine equation*. The problem translates immediately into geometric language. From this point of view view, the problem consists in finding the points on the hypersurface $f(x_1, \cdots, x_n) = 0$ in $A_n(k)$ having integral co-ordinates, where k is the field of rational numbers. We refer the reader to Nagell [12] for further information on this venerable branch of number theory.

It is important to realize at the outset that not all curves are rational. The question of determining the rationality, or non-rationality, of a curve leads into rather deep waters and discussion here is inappropriate. Rather, it seems more desirable in such a preliminary account to study in detail several examples. In this way, some insight will be gained into the geometric situations that arise.

In the following examples, all curves are assumed to be in $A_2(k)$ where k is the field of real numbers; they will be interpreted as curves in the euclidean plane as occasion demands.

Example 1: Any curve of the form $y = f(x)$ where $f(x)$ is a polynomial in x is a rational curve; it can be parametrized by $x = t$, $y = f(t)$. If $f(x)$ has integral coefficients, then all integral points on the curve clearly are obtained by letting

t take on all integral values and so this solves completely the diophantine equation $y - f(x) = 0$.

Example 2: The curve $x^6 - x^2y^3 - y^5 = 0$ is rational. To see this, cut it with the line $y = tx$ where t is a new variable (Fig. 3-1). (Note that every point in $A_2(k)$ is on the line $y = tx$ for some value of t, or else on the line $x = 0$.) For $x \neq 0$, we obtain the point of intersection $(t^3 + t^5, t^4 + t^6)$. This also gives the point (0, 0) if we take $t = 0$. Thus *every* point on the curve is given by the parametric equations

$$(1) \qquad \begin{cases} x = t^3 + t^5 \\ y = t^4 + t^6 \end{cases}$$

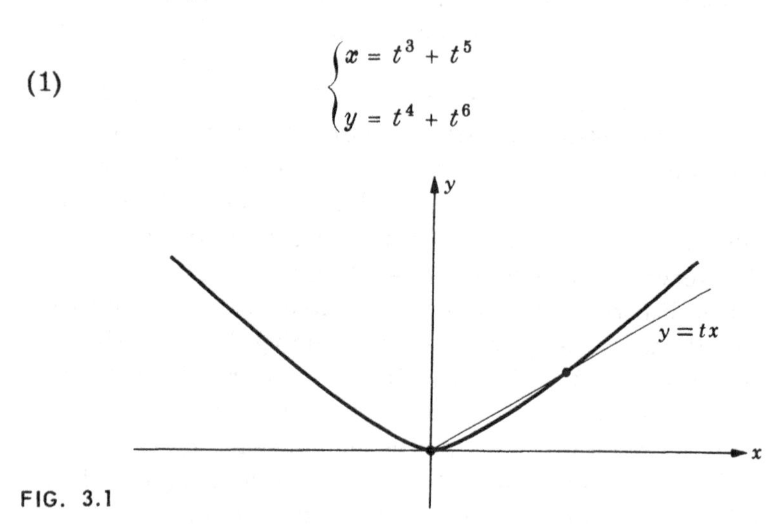

FIG. 3.1

The parametrization (1) enables us to find all the integral points on the curve very easily. Certainly if we take integral values for t in (1) we get integral points on the curve. We contend that these are *all* the integral points on the curve. Suppose (x, y) is an integral point. Then x and $y = tx$ are integers. Since $x = t^3(1 + t^2)$ the only way x can be zero is for t to be zero. Thus it may be assumed that x is not zero and so $t = \dfrac{y}{x}$ is rational. We write t in the form

$t = \dfrac{a}{b}$ where a and b are relatively prime integers, that is, they have no common division other than ± 1. Now $x = t^3 + t^5$ and so $x = \dfrac{a^3b^2 + a^5}{b^5}$. This must be an integer and so $a^3b^2 + a^5 \equiv 0 \pmod{b}$. Therefore $a^5 \equiv 0 \pmod{b}$. Now a and b are relatively prime and so $a \equiv 0 \pmod{b}$. The only way this can happen is to have $b = \pm 1$. Thus t must be an integer. This proves our contention and so we get all solutions to the diophantine equation $x^6 - x^2y^3 - y^5 = 0$ by letting t range over all integral values in the equations (1).

Example 3: The same kind of trick works for the curve $y^2 = x^3 + x^2$ (Fig. 3-2). We cut it with the line $y = tx$ of

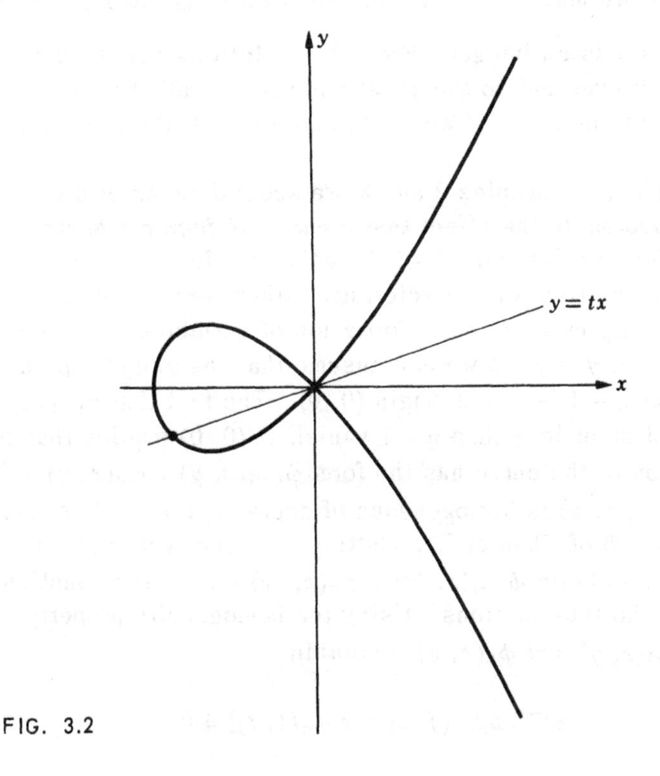

FIG. 3.2

variable slope t to obtain $t^2x^2 = x^3 + x^2$. Thus for $x \neq 0$ we get $x = t^2 - 1$ and $y = t(t^2 - 1)$. Again, the origin is included here if we let $t = \pm 1$. Thus all points on the curve are given by the parametric equations

$$(2) \qquad \begin{cases} x = t^2 - 1, \\ y = t(t^2 - 1). \end{cases}$$

Clearly an infinity of integral points are obtained if we let t range over all integral values in equations (2). Again the converse is true. For suppose (x, y) is an integral point. As before, the case $x = 0$ is obtained from integral values of t, so we may assume $x \neq 0$. Then $t = \dfrac{y}{x}$ is rational and $t^2 = x + 1$ is an integer. From this it follows that t must be an integer and so the equations (2) give all the integral points on the curve if we let t range over all the integers.

Example 4: Examples 2 and 3 are special cases of a general theorem to the effect that *a curve of degree n having a singular point of order n - 1 is rational.* Here we can take k to be any field of characteristic either zero or greater than n. By making a transformation of variables of the form $x' = x - a$, $y' = y - b$ we can assume that the singular point of order $n - 1$ is at the origin $(0, 0)$. The fact that all partials of order less than $n - 1$ vanish at $(0, 0)$ implies that the equation of the curve has the form $\phi_{n-1}(x, y) + \phi_n(x, y) = 0$ where $\phi_l(x, y)$ is homogeneous of degree l; $l = n - 1, n$ (cf. Exercise 5 of Chapter I.). Cutting the curve with the line $y = tx$ we obtain $\phi_{n-1}(x, tx) + \phi_n(x, tx) = 0$ for the equation giving the intersections. Using the homogeneity property of $\phi_{n-1}(x, y)$ and $\phi_n(x, y)$ we obtain

$$x^{n-1}[\phi_{n-1}(1, t) + x\, \phi_n(1, t)] = 0.$$

For $x \neq 0$ this gives

$$(3) \qquad \begin{cases} x = \dfrac{-\phi_{n-1}(1, t)}{\phi_n(1, t)} \\[2ex] y = \dfrac{-t\,\phi_{n-1}(1, t)}{\phi_n(1, t)} \end{cases}$$

provided $\phi_n(1, t) \neq 0$.

Now there are at most a finite number of values of t for which $\phi_n(1, t) = 0$ and the corresponding lines $y = tx$ intersect the curve in, at most, a finite number of points. Also, there are only a finite number of points on the curve with $x = 0$. Thus the parametric equations (3) give all the points on the curve except for, at most, a finite number. It is clear that any value of t for which $\phi_n(1, t) \neq 0$ gives a point on the curve. Thus the curve is rational.

Example 5: In the specific examples we have looked at so far, the parametrizations had no exceptions. One of the simplest cases where exceptions occur is that of the circle $x^2 + y^2 = a^2$ $(a > 0)$. Obviously any line $y = tx$ will cut the curve in two points and so our previous procedure must be modified. What we do is cut the curve with the line $y = t(x + a)$ (Fig. 3-3).

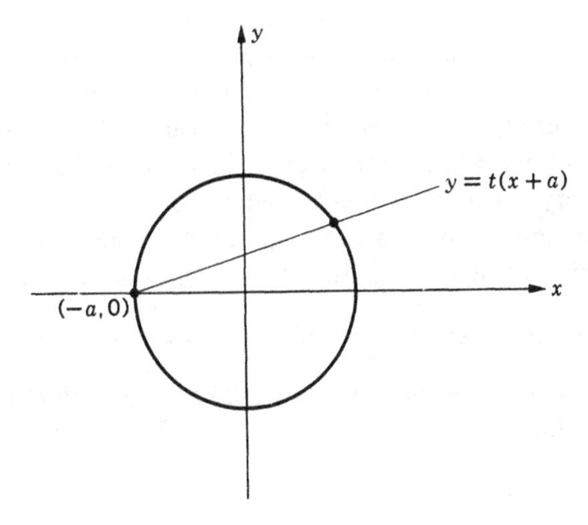

FIG. 3.3

A simple algebraic computation gives for the intersection points $(-a, 0)$ and

$$\left(\frac{a(1-t^2)}{1+t^2}, \frac{2at}{1+t^2}\right).$$

Conversely, this latter point is on the circle for any value of t, as can be verified directly. However, there is no value of t that will make the second point be $(-a, 0)$. Thus we get parametric equations for the circle

$$\text{(4)} \qquad \begin{cases} x = \dfrac{a(1-t^2)}{1+t^2}, \\[2ex] y = \dfrac{2at}{1+t^2}, \end{cases}$$

and the point $(-a, 0)$ is missed. One naturally is annoyed that this should happen with a circle after everything has gone along perfectly well with the previous examples in which singularities of some complexity were present. Let us see what can be done.

First of all, since we are working over the real numbers, we can talk about limits. In particular, the equations (4) allow us to compute the limit of (x, y) as t increases indefinitely. The limit is $(-a, 0)$, the missing point. Thus $(-a, 0)$ corresponds to an "infinite" value of t. This, of course, is a vulgarity that cannot be countenanced in function theory. The standard procedure in such cases is to turn the independent variable upside down and set it (the new variable) equal to zero. This suggests that in (4) we introduce new variables u and v where $t = \dfrac{u}{v}$. It is assumed for the moment that $v \neq 0$. The equations (4) then take the form

$$(5) \qquad \begin{cases} x = \dfrac{a(v^2 - u^2)}{u^2 + v^2}, \\ \\ y = \dfrac{2auv}{u^2 + v^2}. \end{cases}$$

To each ordered pair (u, v) with $v \neq 0$ corresponds a point on the circle, and (su, sv) will determine the same point if s is a non-zero real number. If we now drop the restriction $v \neq 0$, we see that the ordered pair $(1, 0)$, indeed $(s, 0)$ for any real number $s \neq 0$, corresponds to the point $(-a, 0)$. This situation finds its most appropriate setting in the language of projective spaces. What we have constructed is a mapping of points (u, v) of $P_1(k)$ onto the circle; indeed we have a one-to-one correspondence between $P_1(k)$ and the circle in $A_2(k)$ where the mapping $(u, v) \longrightarrow (x, y)$ is given by rational functions. Aside from the important fact of having constructed a new model of real projective 1-space, the principal moral of this example is that to obtain parametrizations of rational curves in which no exceptional points are missed, it apparently is necessary to extend consideration to projective spaces, even though the original curve lies in the affine plane. Indeed, in the present case, the projective curve obtained by "homogenizing" the equation of the circle has no points at infinity at all; it lies entirely in the embedded affine space.

Example 6: We have seen in the previous example that in order to take care of certain exceptional points in a parametrization it is necessary to introduce projective spaces. This does not resolve the difficulty in full; the field k may be too small. To see how this can happen, we shall consider the particular curve $y^2 = x^4 - x^2$ (Chapter I, Example 3). We begin by cutting it with the line $y = tx$ to obtain

$x^2(1 + t^2) = x^4$ for the equation giving the intersections (Fig. 3-4).

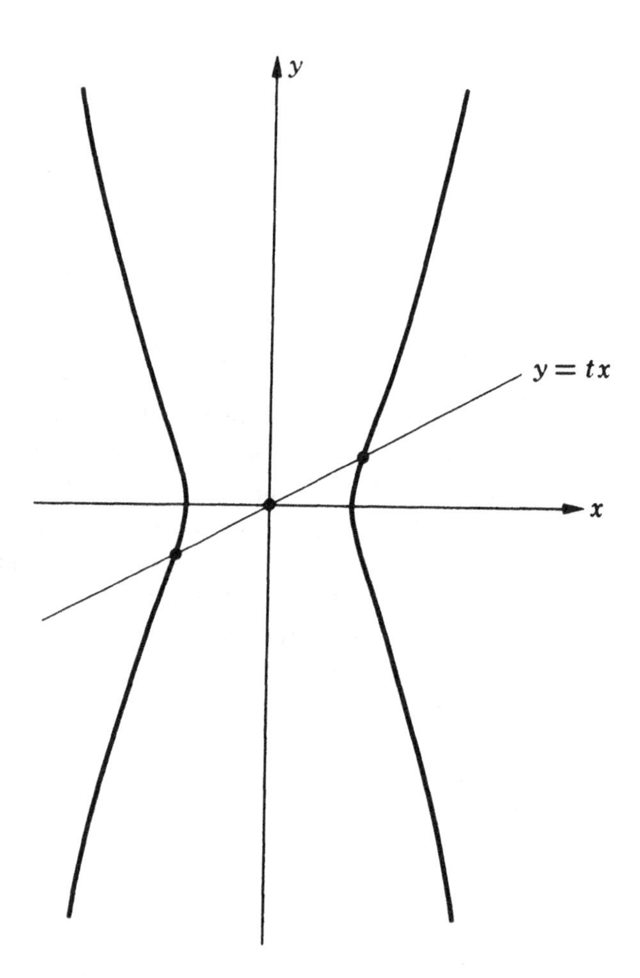

FIG. 3.4

If $x \neq 0$ we obtain $x^2 - t^2 = 1$. Unfortunately, for a given value of t this equation will have, in general, *two* solutions. However, the equation $x^2 - t^2 = 1$ can be considered as the equation of a hyperbola, and this can be rationally parametrized by cutting it with the line $t = u(x - 1)$ of variable slope u (Fig. 3-5).

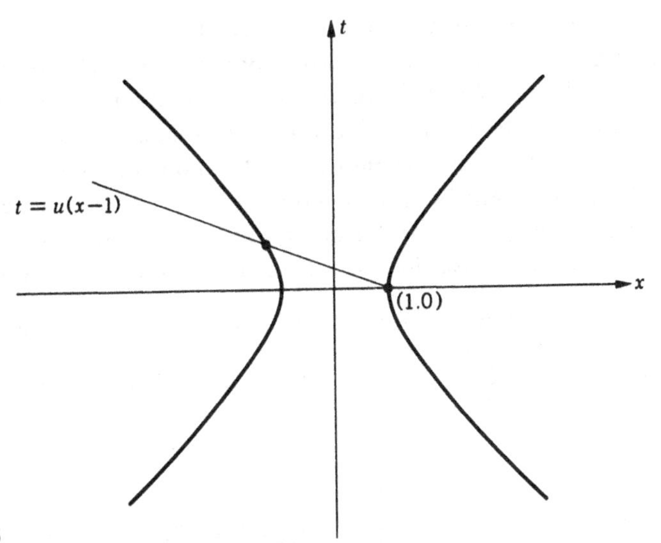

FIG. 3.5

For the intersections of this line with the hyperbola we obtain (1, 0) and $\left(\dfrac{u^2+1}{u^2-1}, \dfrac{2u}{u^2-1}\right)$. Conversely, this latter point is on the hyperbola for any value of $u \neq \pm 1$. Thus the hyperbola has parametric equations

$$\text{(6)} \qquad \begin{cases} x = \dfrac{u^2+1}{u^2-1}, \\[2ex] t = \dfrac{2u}{u^2-1}. \end{cases}$$

This gives all points on the hyperbola except the point (1, 0). Note that the values $u = \pm 1$ do not correspond to any point on the curve. Going back now to the original curve, using $y = tx$, we obtain as parametric equations

$$\text{(7)} \qquad \begin{cases} x = \dfrac{u^2+1}{u^2-1}, \\[2ex] y = \dfrac{2u(u^2+1)}{(u^2-1)^2}. \end{cases}$$

The only points that might have been missed are the origin (0, 0) and the point corresponding to (1, 0) on the hyperbola. The first of these is indeed missed since no value can be assigned to u that will make $x = 0$. The point on the curve corresponding to (1, 0) on the hyperbola is (1, 0) and this is missed since $u^2 + 1 > u^2 - 1$ for all u. The point (1, 0) can be taken care of in much the same way as was used for $(-a, 0)$ in the case of the circle. We introduce new variables r and s with $u = \dfrac{r}{s}$, assuming for the moment that $s \neq 0$. The equations (7) become

$$(8) \qquad \begin{cases} x = \dfrac{r^2 + s^2}{r^2 - s^2}, \\[2ex] y = \dfrac{2rs(r^2 + s^2)}{(r^2 - s^2)^2}. \end{cases}$$

This gives a mapping $(r, s) \longrightarrow (x, y)$ from real projective 1-space to the curve, which is defined for all points in $P_1(k)$ except (1, –1) and (1, 1). We certainly obtain all the points we had before, together with one new one, namely (1, 0). This is obtained by taking $r = 1$ and $s = 0$. Unfortunately we still do not obtain the point (0, 0). For this, the only possibility appears to be that we must let the parameters take on complex values. Then $r = \pm i$, $s = 1$ will give the point (0, 0). This is another illustration of the desirability of working over an algebraically closed field or, at any rate, one which is sufficiently large.

One further remark remains to be made. The mapping from $P_1(k)$ to the curve is not defined at (1, 1) and (1, –1). This unpleasantness is removed by taking the projective curve corresponding to $y^2 = x^4 - x^2$. We set $x = \dfrac{x_1}{x_3}$ and $y = \dfrac{x_2}{x_3}$ and obtain for the corresponding projective curve the equation $x_2^2x_3^2 = x_1^4 - x_1^2x_3^2$. Then the parametric equations can be written

$$(9) \qquad \begin{cases} x_1 = (r^2 + s^2)(r^2 - s^2), \\ x_2 = 2rs(r^2 + s^2), \\ x_3 = (r^2 - s^2)^2. \end{cases}$$

For $r^2 \neq s^2$, this amounts to the same thing as equations (8). The only new point we get is (0, 1, 0), and that by taking $r = \pm s$. This new point is the single point at infinity on the projective curve. Thus equations (9) give a mapping from $P_1(k)$ to the projective curve which is defined for *all* points of $P_1(k)$. In order to get all points of the curve as images of points of $P_1(k)$, it is necessary to adjoin the complex unit i to the real numbers. Thus if k is the field of complex numbers, the mapping in question is everywhere defined and maps $P_1(k)$ onto the whole projective curve.

Examples of rational algebraic sets of higher dimension will be considered in Chapter V.

Exercises for Chapter III

1. In Example 3 what part of the curve corresponds to the range $0 \leq t < +\infty$; to $-\infty < t \leq 0$?
2. Show that the problem of finding the rational points on the circle $x^2 + y^2 = 1$ is equivalent to finding the integral points on the surface $x^2 + y^2 = z^2$. Using the parametrization of Example 5 with $a = 1$, find all integral points on this surface.
3. In Example 5 a one-to-one correspondence is constructed between the circle and real projective 1-space. How does this agree, topologically, with the description of $P_1(k)$, k = field of real numbers, as the circle with antipodal points identified?

4. Suppose C is a curve of degree n with a singular point of order $n - 1$ at $(0, 0)$. Also, suppose that k is the field of complex numbers and that the equation of C does not factor. Extend the parametrization (3) of Example 4 to the corresponding projective curve to obtain $x_1 = \phi_{n-1}(t)$, $x_2 = t\,\phi_{n-1}(t)$, $x_3 = \phi_n(1, t)$ and thence a mapping of $P_1(k)$ onto the whole projective curve. (Hint: The fact that the equation of C does not factor prohibits $\phi_{n-1}(1, t)$ and $\phi_n(1, t)$ from having a common zero.)
5. Show that in equations (9) we cannot have $x_1 = x_2 = x_3 = 0$ for $(r, s) \in P_1(k)$.
6. Show that the following curves are rational and enumerate exceptional points.

(i) $y^2 = \dfrac{x^2(a + x)}{a - x}$,

(ii) $y^2 = \dfrac{x^3}{a - x}$,

(iii) $x^3 + y^3 + 3axy = 0$,

(iv) $y^2 = \dfrac{x^2}{(x + 1)(x + 2)}$, (Hint: Proceed as in example 6.)

(v) $y^2 = x^3 - x^2$.

IV

ALGEBRAIC SETS WITH GROUP STRUCTURE

Among the most interesting kinds of mathematical objects are those in which different types of mathematical structures are combined. The prototype of this situation is the real numbers. From the purely algebraic point of view they form an abelian group under addition; from a geometrical point of view they can be considered as constituting a one-dimensional affine space; also there is a notion of limit so one can do topology too. These three aspects of the real numbers have to be compatible in some appropriate technical sense; for instance, to mention just one example, the function $f(x) = x + a$ must be continuous. All function theory stems ultimately from the interplay of these various structures on the real numbers. Readers familiar with such things as Lie groups and Banach spaces will recognize the same sort of situation there.

In this chapter we shall be concerned with mathematical objects which combine the structure of algebraic set with that of a group. The theory of these so-called *algebraic groups* is one of the most important recent achievements in algebraic geometry. It includes as a special case the modern theory of abelian varieties. Obviously it would be an impertinence to attempt here anything like the beginning of a general theory; and it would be pointless to do so without the appropriate algebraic and geometric machinery. As in the last chapter we shall try to convey the underlying idea by detailed consideration of several elementary examples. The general idea in all these examples is that of an algebraic set V for which a notion of multiplication of points is defined so that V forms a group. We also require that the group structure on V be compatible with the geometric structure of V in the sense that if p and q are points of V with co-ordinates $(x_1, \cdots, x_n)$ and $(y_1, \cdots, y_n)$ respectively, then the co-ordinates of $p \cdot q$ are rational functions of $x_1, \cdots, x_n$ and $y_1, \cdots, y_n$ and the co-ordinates of p^{-1} are rational functions of $x_1, \cdots, x_n$.

Example 1: The simplest example to start with is $A_n(k)$ where k is an arbitrary field. If $p = (x_1, \cdots, x_n)$ and $q = (y_1, \cdots, y_n)$ we can define $p \cdot q$ by

$$p \cdot q = (x_1 + y_1, \cdots, x_n + y_n).$$

Now $A_n(k)$ is certainly an algebraic set — the locus of any equation which is an identity. The group operations are given by rational functions, indeed polynomial functions, and $A_n(k)$ forms an abelian group under this multiplication.

Example 2: Suppose V is a hypersurface in $A_n(k)$ given by an equation $x_n = f(x_1, \cdots, x_{n-1})$ where $f(x_1, \cdots, x_{n-1})$ is a polynomial with coefficients in k. This is a rational hypersurface with the agreeable property that there is a one-to-one correspondence between its points and $(n - 1)$-tuples $(c_1, \cdots, c_{n-1})$ given by the mapping

$$(c_1, \cdots, c_{n-1}) \longrightarrow (c_1, \cdots, c_{n-1}, f(c_1, \cdots, c_{n-1}))$$

of $A_{n-1}(k) \longrightarrow V$. The multiplication of points of $A_{n-1}(k)$ induces a multiplication of points on V and makes it into a group. The group operations on V are again given by rational functions. The hypothesis on the form of $f(x_1, \cdots, x_{n-1})$ can be weakened to assuming that it is a rational function with a denominator which never vanishes for values in k. For instance, our discussion applies to the curve $y = \frac{1}{1 + x^2}$ in $A_2(k)$ where k is a subfield of the real numbers.

Example 3: Let C be the circle $x^2 + y^2 = 1$ in $A_2(k)$ where k is the field of real numbers. We shall interpret this as a curve in the euclidean plane so that certain helpful implements, namely the trigonometric functions, will be available. We parametrize C by the equations $x = \cos\theta$, $y = \sin\theta$ where θ is the arc length measured counterclockwise along the circumference from the point (1, 0). Suppose $p_1 = (\cos\theta_1, \sin\theta_1)$ and $p_2 = (\cos\theta_2, \sin\theta_2)$ are any two points on C. We define $p_1 \cdot p_2$ to be the point $(\cos(\theta_1 + \theta_2), \sin(\theta_1 + \theta_2))$. This gives C a structure of abelian group. If the co-ordinates of p_1 and p_2 are (x_1, y_1) and (x_2, y_2) respectively, then from the addition formulas of trigonometry it follows easily that

$$(1) \qquad p_1 \cdot p_2 = (x_1 x_2 - y_1 y_2, x_1 y_2 + x_2 y_1)$$

and so the co-ordinates of the product are rational functions, indeed polynomial functions, of the co-ordinates of the "factors"; this is also true for the co-ordinates of p_1^{-1}.

Example 4: The equation (1) of Example 3 can be interpreted to give a multiplication on $P_1(k)$ where k is any subfield of the real numbers.

If $p_1 = (x_1, y_1)$ and $p_2 = (x_2, y_2)$ are two points of $P_1(k)$ written in terms of homogeneous co-ordinates, we define $p_1 \cdot p_2$ by equation (1). It is important to note that if (x_1, y_1) is replaced by (tx_1, ty_1) with $t \neq 0$ in k, the equation (1) gives the same product point in $P_1(k)$; this is also true if (x_2, y_2) is replaced by (tx_2, ty_2) with $t \neq 0$ in k. Furthermore the product point is in $P_1(k)$ since the sum of the squares of its co-ordinates is non-zero. (This is where we use the fact that k is a subfield of the real numbers.) Thus the multiplication of points on $P_1(k)$ is well defined and it is easy to verify that the group axioms are satisfied. The group operations are, of course, given by homogeneous polynomial functions.

Example 5: The previous example of multiplication on the circle can be generalized to higher dimensions. Let k be the field of real numbers and let $\theta_1, \cdots, \theta_n$ be independent variables. The points of the form

$$(\cos\theta_1, \sin\theta_1, \cos\theta_2, \sin\theta_2, \cdots, \cos\theta_n, \sin\theta_n)$$

constitute an algebraic set V in $A_{2n}(k)$. Indeed if we write the co-ordinates of points in the form

$$(x_1, y_1, x_2, y_2, \cdots, x_n y_n)$$

then this algebraic set is exactly that defined by the equations $x_i^2 + y_i^2 = 1$ for $i = 1, 2, \cdots, n$. If $p_1 = (\cdots, x_i^{(1)}, y_i^{(1)}, \cdots)$ and $p_2 = (\cdots, x_i^{(2)}, y_i^{(2)}, \cdots)$ are any two points of V, then we define $p_1 \cdot p_2$ to be the point

$$(\cdots, x_i^{(1)}x_i^{(2)} - y_i^{(1)}y_i^{(2)}, x_i^{(1)}y_i^{(2)} + x_i^{(2)}y_i^{(1)}, \cdots).$$

This gives V the structure of group and as such, V is isomorphic to a direct product of circle groups.

Example 6: *Algebraic matrix groups*. The ring $[k]_n$ of matrices of degree n over a field k can be thought of as the affine space $A_{n^2}(k)$ by regarding the coefficients of a matrix as "co-ordinates" of a point. To be quite specific, we can write the rows of a matrix one *after* another instead of below; for instance we can think of the *matrix*

$\begin{pmatrix} a_{11} & a_{12} \\ a_{21} & a_{22} \end{pmatrix}$ as being the *point* $(a_{11}, a_{12}, a_{21}, a_{22})$. Consider first of all the set of matrices $X \in [k]_n$ such that ${}^tX.X = I$ where, as usual, the pre-superscript denotes the transpose and I denotes the identity matrix. The set of all such matrices forms a group; in case k is the field of real numbers, we obtain the real *orthogonal group of degree n.* The defining condition amounts to certain polynomial equations on the coefficients of X, and so this group is an algebraic set in $A_{n^2}(k)$. Also the group operations are given by polynomial functions of the coefficients. A similar situation prevails for the *symplectic group* $Sp_{2n}(k)$. This group is defined over *any* field as follows: let $E = \begin{pmatrix} 0 & I \\ -I & 0 \end{pmatrix}$ where I is the identity matrix of degree n. Then $Sp_{2n}(k)$ consists of the set of all matrices $X \in [k]_{2n}$ such that ${}^tXEX = E$. Once again, the group obtained appears geometrically as an algebraic set in $A_{(2n)^2}(k)$ and the group operations are given by polynomial functions.

Remark: We omit the proof that the examples considered here are groups. To accomplish this, it is much easier to start from another point of view, namely that of linear transformations keeping invariant certain bilinear forms. The reader is referred to any of the standard works on linear algebra, for example, Bourbaki [3].

Other groups for which a similar situation prevails are the *special linear group* $SL_n(k)$ which consists of those matrices $X \in [k]_n$ such that $\det X = 1$, and the real *Lorentz group* which consists of the matrices $X \in [k]_4$ such that ${}^tXFX = F$ where k is the field of real numbers and F is the diagonal matrix $\begin{pmatrix} 1 & 0 & 0 & 0 \\ 0 & 1 & 0 & 0 \\ 0 & 0 & 1 & 0 \\ 0 & 0 & 0 & -1 \end{pmatrix}$.

For the *full linear group* $GL_n(k)$, an important difference arises. This group consists of the invertible matrices in $[k]_n$ where k is an arbitrary field. Thus the condition for a matrix X to belong to this group is that $\det X \neq 0$. Now this does *not* give polynomial *equations* for the coefficients and so $GL_n(k)$ is not an algebraic set as it stands, but rather consists of an algebraic set $[k]_n$, with a certain algebraic subset — namely, that defined by $\det X = 0$ — deleted.

It is nevertheless possible to discuss this group from a geometric point of view, and similar situations have partly been the motivation for the extensive generalizations of the notion of algebraic set. However, so far as the present case is concerned, it is possible to embed $GL_n(k)$ as an algebraic set in $A_{n^2+1}(k)$. This is done as follows: to the variables x_{ij}, the coefficients of the matrix X, we add a new variable t, and regard the x_{ij} and t as co-ordinates of a point in $A_{n^2+1}(k)$. Then $GL_n(k)$ is in one-to-one correspondence with the algebraic set V in $A_{n^2+1}(k)$ defined by $t \cdot \det(x_{ij}) = 1$, a matrix (a_{ij}) of $GL_n(k)$ corresponding to the point (a_{ij}, t_a) of V where $t_a = [\det(a_{ij})]^{-1}$. It is easy to check that this gives a one-to-one correspondence between $GL_n(k)$ and V.

Example 7: Multiplication on a non-singular cubic curve. Let V be the curve

$$x_2^2 x_3 = x_1(x_1 - a_1 x_3)(x_1 - a_2 x_3)$$

in $P_2(C)$ where C is the field of complex numbers and a_1, a_2 are distinct non-zero elements of C. This is the projective form of the (affine) elliptic curve $y^2 = x(x - a_1)(x - a_2)$. The curve V has no singular points (cf. problem 3 of Chapter II) and so there is at each point of V a unique tangent. It is possible to define a multiplication of points on V. For any two points p and q on V let $L_{p,q}$ denote the line joining p and q if $p \neq q$, and let it denote the tangent at p if $p = q$. Now any line cuts V in exactly three points if intersections are counted with their proper multiplicity. Thus the line $L_{p,q}$ will cut V in a third point r.

Now let e be the particular point $(0, 0, 1)$. Then the line $L_{e,r}$ cuts V in a third point which we shall call the product of p and q and denote by $p \circ q$. In this way a group structure is defined on V. Notice that $p \circ q = q \circ q$ for any $p, q \in V$ and so the group is commutative. The construction is illustrated by Fig. 4-1 which represents the real part of the curve in the affine plane for this case $0 < a_1 < a_2$.

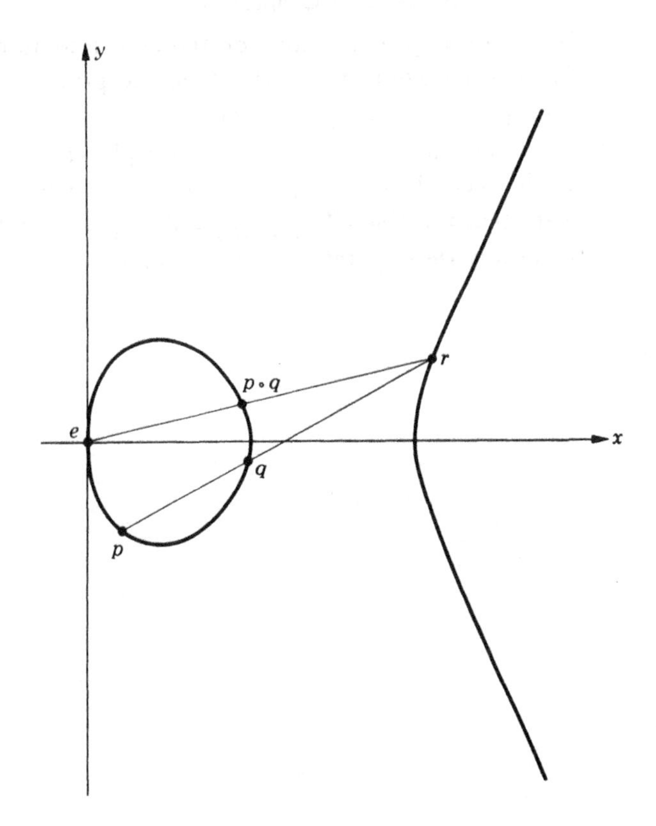

FIG. 4.1

The point e acts as the identity element of the group. With one exception, it is easy to check that the group axioms are satisfied. The exception is associativity, which is rather difficult to establish with the methods we have developed so far. (This, however, should not discourage the attempt.)

This example is one of the simplest cases of an *abelian variety* (cf. Lang [10]) and is intimately connected with the theory of elliptic functions. The sketch given here is admittedly but the barest outline. For further details the reader is referred to van der Waerden [14] and Walker [16].

Exercises for Chapter IV

1. Write down the conditions on the coefficients of a matrix for it to belong to the real orthogonal group of degree n; the symplectic group $Sp_{2n}(k)$.
2. Show that the associativity of multiplication on the cubic curve V of Example 7 is equivalent to the statement that the lines $L_{p_1, p_2 \circ p_3}$ and $L_{p_1 \circ p_2, p_3}$ intersect on V for any three points p_1, p_2, p_3 of V.

V

SEGRE AND VERONESE VARIETIES

1. Product structures

The construction of direct products is a procedure which, under various guises, appears in many different parts of mathematics. To take the simplest instance, let A and B be any two sets. The set of ordered pairs (x, y) with $x \in A$ and $y \in B$ is called the *cartesian product* of A and B and is denoted $A \times B$. Very often the two sets A and B are endowed with some kind of mathematical structure and we wish to impose a similar structure on the set $A \times B$ so that the components A and B are embedded in $A \times B$ in some suitable manner. For instance, if A and B are groups we can define multiplication in $A \times B$ component-wise so that it becomes a group, the so called *direct product* of A and B.

Another situation that arises is that in which the sets A and B are topological spaces. The set $A \times B$ can then be made into a topological space, the *product space*. This construction appears very frequently in topology and analysis. For instance, the euclidean plane appears from this point of view as the product of the real line with itself. The construction of products arises also in algebraic geometry and plays a part no less prominent than it does in algebra and topology. The notion of products is particularly important in general intersection theory and in the theory of correspondences. We now proceed to consider the fundamental ideas involved.

2. Products of affine and projective spaces, Segre varieties

The product of two affine spaces is constructed in a very simple manner. Let $A_m(k)$ and $A_n(k)$ be the affine spaces of dimension m and n respectively over an arbitrary field k. The cartesian product of these two spaces, as sets, consists of the set of ordered pairs (p, q) where $p \in A_m(k)$ and $q \in A_n(k)$. This cartesian product can be given the structure of an affine space of dimension $m + n$. Indeed if $p = (x_1, \cdots, x_m)$ and $q = (y_1, \cdots, y_n)$, the mapping

$$\phi: (p, q) \longrightarrow (x_1, \cdots, x_m, y_1, \cdots, y_n)$$

is a one-to-one mapping of $A_m(k) \times A_n(k)$ *onto* $A_{m+n}(k)$. Since the x_i and y_i can be chosen arbitrarily in k, every point of $A_{m+n}(k)$ is the image of some element (p, q). If V_m is an algebraic set in $A_m(k)$ and V_n an algebraic set in $A_n(k)$ then $\phi(V_m \times V_n)$ is clearly an algebraic set in $A_{m+n}(k)$ (cf. Exercise 1).

Now, suppose we try the same sort of procedure on projective spaces $P_m(k)$ and $P_n(k)$. First of all, we can construct the cartesian product $P_m(k) \times P_n(k)$ in the purely set-theoretical sense. The attempt to set up a mapping analogous to the ϕ used in the affine case breaks down. To see why this is so, let $p = (x_0, \cdots, x_m) \in P_m(k)$ and $q = (y_0, \cdots, y_n) \in P_n(k)$. We would try to map (p, q) into the point $(x_0, \cdots, x_m, y_0, \cdots, y_n)$ in $P_{m+n+1}(k)$. However, this mapping is not well-defined. The co-ordinates of p can be multiplied by any non-zero constant and so can those of q. In other words if λ, μ are non-zero elements of k then we would have to have

$$(\lambda x_0, \cdots, \lambda x_m, \mu y_0, \cdots, \mu y_n) = (\nu x_0, \cdots, \nu x_m, \nu y_0, \cdots, \nu y_n)$$

for some $\nu \neq 0$ in k. Since some x_i is non-zero, and also some y_j, this would necessitate the conclusion that $\nu = \lambda = \mu$. Now of course it is perfectly reasonable to have chosen λ different from μ and this would have intolerable consequences. In other words, the "alleged" mapping depends on the choice of the homogeneous co-ordinates of the points p and q, not on the points themselves. Even had the mapping worked, it would not have been surjective onto $P_{m+n+1}(k)$ since each image point would have had at least *two* non-zero co-ordinates. What we do now is set $z_{ij} = x_i y_j$ for $i = 0, \cdots, m$ and $j = 0, \cdots, n$. Since some x_i and some y_j are non-zero, it follows that some z_{ij} is non-zero. Consequently we can regard the z_{ij} as homogeneous co-ordinates of a point in $P_N(k)$ where $N = (m + 1)(n + 1) - 1$. It is understood that these co-ordinates are written out in some fixed order. For instance we could use lexicographical order which in the case $m = 1$, $n = 2$ would look as follows: $(z_{00}, z_{01}, z_{02}, z_{10}, z_{11}, z_{12})$. With this understanding, we denote the point with co-ordinates z_{ij} by (z_{ij}).

We contend that the mapping

$$\phi: (p, q) \longrightarrow (x_i y_j) \tag{1}$$

is a one-to-one mapping of $P_m(k) \times P_n(k)$ into $P_N(k)$. First of all, it is well-defined. If the x_i are replaced by λx_i and the y_j by μy_j where λ, μ are non-zero elements of k, then $x_i y_j$ is replaced by $\lambda\mu x_i y_j$, which gives the same point in $P_N(k)$.

We now show that ϕ is one-to-one. Suppose we are given a point (z_{ij}) in the image. Then $z_{ij} = x_i y_j$ and we have to show that the *points* $(x_0, \cdots, x_m)$ and $(y_0, \cdots, y_n)$ are uniquely determined. We know that some y_k is non-zero. Suppose $y_k \neq 0$. Then as *points*, $(x_0, \cdots, x_m)$

$= (x_0 y_k, \cdots, x_m y_k) = (z_{0k}, \cdots, z_{mk})$. Similarly some $x_l \neq 0$ and as *points*, $(y_0, \cdots, y_n) = (x_l y_0, \cdots, x_l y_n)$ $= (z_{l0}, \cdots, z_{ln})$. Thus for a given point in the image, the element (p, q) from which it comes is uniquely reconstructible and so ϕ is one-to-one.

The mapping ϕ is *not* surjective. If (z_{ij}) is an image point, it is immediately verified by taking $z_{ij} = x_i y_j$, that it satisfies the system of $\binom{m+1}{2} \cdot \binom{n+1}{2}$ equations

$$z_{ik} z_{jl} = z_{il} z_{jk} \quad \text{for} \quad i \neq j \quad \text{and} \quad k \neq l. \tag{2}$$

It is trivial to produce values of the z_{ij}, naturally not of the form $x_i y_j$, such that these equations are not satisfied. Thus there are points of $P_N(k)$ which do not occur as images under ϕ. The equations (2) determine a certain algebraic set $V_{m,n}$ in $P_N(k)$ of which it is a proper subset.

Theorem: The mapping $\phi: P_m(k) \times P_n(k) \longrightarrow V_{m,n}$ given by $\phi: (p, q) \longrightarrow (x_i y_j)$ is surjective.

Proof: We have to show that $V_{m,n}$ consists exactly of the points (z_{ij}) where $z_{ij} = x_i y_j$. In other words, given a point (c_{ij}) on $V_{m,n}$ we have to produce a point $(a_0, \cdots, a_m)$ in $P_m(k)$ and a point $(b_0, \cdots, b_n)$ in $P_n(k)$ such that $c_{ij} = a_i b_j$. Suppose (c_{ij}) is a point of $V_{m,n}$. At least one of the co-ordinates of this point must be non-zero, so assume $c_{st} \neq 0$. We first suppose that the required a_i and b_j can be found. Then $c_{sj} = a_s b_j$ for $j = 0, 1, \cdots, n$. Take $a_s = 1$. Then $b_j = c_{sj}$. On the other hand, $c_{it} = a_i b_t$ for $i = 0, 1, \cdots, m$. Now $b_t = c_{st} \neq 0$ and so $a_i = b_t^{-1} c_{it}$ $= c_{st}^{-1} c_{it}$. (Note that this is consistent with the previous choice $a_s = 1$.) To summarize so far, we take $a_i = c_{st}^{-1} c_{it}$ and $b_j = c_{sj}$. We have to show that $c_{ij} = a_i b_j$ for all i, j.

Now $a_i b_j = c_{st}^{-1} c_{it} c_{sj}$. If $i = s$ or if $t = j$ then $a_i b_j = c_{ij}$. So we may assume $i \neq s$ and $t \neq j$. On using the equations (2), it follows that $c_{it} c_{sj} = c_{ij} c_{st}$ and so

$$a_i b_j = c_{st}^{-1} c_{it} c_{sj} = c_{st}^{-1} c_{ij} c_{st} = c_{ij}.$$

This completes the proof of the theorem.

Definition: The algebraic set $V_{m,n}$ given by the equations (2) is called the *Segre variety** associated with the product $P_m(k) \times P_n(k)$.

It is important to note that a Segre variety is *rational* in the sense that the co-ordinates of all its points are given by polynomials in a certain number of variables. In this case $z_{ij} = x_i y_j$ where the independent variables are $x_0, \cdots, x_m$ and $y_0, \cdots, y_n$. This concept of rationality is in obvious accord with its interpretation in Chapter 3 for curves in projective 2-space. Note, in particular, that the parametrization has no exceptions.

Example: The simplest case to consider is that in which $m = n = 1$. In this case, the system (2) reduces to one equation $z_{00} z_{11} = z_{01} z_{10}$. This is the equation of a second-degree surface, a quadric surface, in $P_3(k)$. Now suppose that k is the field of complex numbers. Then any non-singular quadric is equivalent by a projective transformation to the quadric $\sum_{i=0}^{3} x_i^2 = 0$ (cf. Chapter II, §9).

*This algebraic set satisfies the extra algebraic conditions (absolute irreducibility) to qualify for the title "variety," a fact that follows easily from the general theory of algebraic varieties (cf. Lang [10]). We do not have here the appropriate algebraic machinery to prove this and do not make any use of this fact in the sequel. The purpose of this note is merely to justify not calling $V_{m,n}$ the "Segre algebraic set," which would sound rather odd to any professional geometer.

We now set

$$\begin{cases} z_{00} = x_0 + ix_1 \\ z_{11} = x_0 - ix_1 \\ z_{01} = -(x_2 + ix_3) \\ z_{10} = x_2 - ix_3 \end{cases}$$

which amounts to a projective transformation in $P_3(k)$ (cf. Exercise 3). Then the given quadric becomes $z_{00} z_{11} = z_{01} z_{10}$. Thus a non-singular quadric in $P_3(k)$, when k is the field of complex numbers, is projectively equivalent to the Segre variety $V_{1,\,1}$. We note one important consequence of this, namely the fact that every such quadric is a rational surface.

3. Veronese varieties

In the particular case $n = m$, the product $P_m(k) \times P_m(k)$ has as subset the *diagonal subset* D. This consists of all pairs (p, p) where $p \in P_m(k)$. The mapping ϕ maps D into a certain subset of the Segre variety $V_{m,\,m}$. If $p = (x_0, \cdots, x_m)$ then ϕ: $(p, p) \longrightarrow (z_{ij})$ where $z_{ij} = x_i x_j$. Thus the points of $\phi(D)$ satisfy the equations

$$(3) \qquad \begin{cases} z_{ik} z_{jl} = z_{il} z_{jk} \quad \text{for} \quad i \neq j \quad \text{and} \quad k \neq l, \\ z_{ij} = z_{ji} \quad \text{for} \quad i, j = 0, 1, \cdots, m. \end{cases}$$

These equations determine an algebraic subset W_m of $V_{m,\,m}$ and this algebraic subset is called the *Veronese variety*[†] associated with the product $P_m(k) \times P_m(k)$.

Theorem: The mapping ϕ: $D \longrightarrow W_m$ is surjective.
Proof: Since $W_m \subset V_{m,\,m}$ it follows that every point (c_{ij}) of W_m has the form $c_{ij} = a_i b_j$ where $(a_0, \cdots, a_m)$ and

[†] The term "variety" is justified here for the same reason as in the case of the Segre variety.

$(b_0, \cdots, b_m)$ are in $P_m(k)$. Now (c_{ij}) must satisfy the second of the equations (3) and so $a_i b_j = a_j b_i$ for all i, j. Some $b_t \neq 0$ and so $a_i = a_t b_t^{-1} b_i$ for all i. Set $\lambda = a_t b_t^{-1}$. Then $\lambda \neq 0$ since some $a_i \neq 0$, and so $a_i = \lambda b_i$ for all i. Thus $(a_0, \cdots, a_m) = (b_0, \cdots, b_m)$ as points of $P_m(k)$ and so the restriction of ϕ to D is surjective.

It follows from this theorem that the Veronese variety W_m is rational since every point on it is of the form $(x_i x_j)$ where $x_i \in k$.

Example: We saw at the end of §2 that any non-singular quadric in $P_3(k)$ where k is the field of complex numbers, is projectively equivalent to the Segre variety $V_{1,1}$ given by the single equation $z_{00} z_{11} = z_{01} z_{10}$. The Veronese variety is determined by this equation and the extra equation $z_{01} = z_{10}$. We now set

$$\begin{cases} x = z_{00}, \\ y = z_{11}, \\ u = z_{01} - z_{10}, \\ v = z_{10} \end{cases}$$

which amounts to a projective transformation of $P_3(k)$. The equations of W_1 now take the form

$$\begin{cases} xy = v^2, \\ u = 0. \end{cases}$$

Thus W_1 appears as a non-singular conic in the plane $u = 0$. All non-singular conics in this plane are projectively equivalent since we are working over the complex numbers. Also, any projective transformation of this plane can be extended to all of $P_3(k)$ by keeping u fixed. Thus any non-singular conic in any plane of $P_3(k)$ is projectively equivalent to the Veronese variety W_1. As in the case of the Segre variety $V_{1,1}$, it follows that every non-singular conic in $P_2(k)$ is a rational curve, without exceptional points.

4. Linear subspaces in Segre varieties

As in §2, let $V_{m,n}$ be the Segre variety associated with $P_m(k) \times P_n(k)$. Points $(x_0, \cdots, x_m)$ of $P_m(k)$ will be abbreviated (x_i), similarly for points of $P_n(k)$. Now let $q = (a_i)$ be a fixed point of $P_n(k)$ and denote by $[q]$ the subset of $P_n(k)$ consisting of the single point q. Then ϕ induces a mapping (one-to-one) of $P_m(k) \times [q]$ into $V_{m,n}$ by $\phi: ((x_i), (a_i)) \longrightarrow (x_i a_j) \in P_N(k)$. If $r = (b_i)$ is another point of $P_n(k)$, then ϕ induces a similar mapping $\phi: ((x_i), (b_i)) \longrightarrow (x_i b_j)$ of $P_m(k) \times [r]$ into $V_{m,n}$. If λ, μ are elements of k, not both zero, then for any two points (x_i) and (x_i') of $P_m(k)$,

$$\phi: ((\lambda x_i + \mu x_i'), (a_i)) \longrightarrow ((\lambda x_i + \mu x_i') \cdot a_j)$$

and

$$((\lambda x_i + \mu x_i') \cdot a_j) = (\lambda x_i a_j) + (\mu x_i' a_j).$$

Thus the line joining any two points (x_i) and (x_i') of $P_m(k)$ maps into the line joining their images in $P_N(k)$. Since ϕ is one-to-one, the image points are distinct if the original points are. Thus ϕ is a one-to-one mapping of $P_m(k) \times [q]$ into $V_{m,n}$ which preserves lines — in the sense just explained — and so the composite mapping $(x_i) \longrightarrow ((x_i), (a_i)) \longrightarrow (x_i a_j)$ can be regarded as an embedding of $P_m(k)$ in $V_{m,n}$ which preserves lines. It is customary to express this state of affairs by saying that $V_{m,n}$ contains $P_m(k)$ — or more correctly its image under the composite mapping above — as a *linear subspace*.

We undoubtedly should clarify a matter here that may cause confusion. The image of $P_m(k)$ in $V_{m,n}$ is linear in the sense that if two points are in the image, then the line joining them is in the image. This does not quite coincide with the meaning of the term linear subspace as used in linear algebra. More precisely, if we use the vector space model of $P_m(k)$ as the set of one-dimensional subspaces of the vector space V_{m+1} of $(m+1)$-tuples over k, then ϕ amounts to a particular one-to-one linear mapping of V_{m+1} into V_{N+1} in the usual sense of linear algebra, *with the exception that we do not apply it to the zero-vector*, since the latter does not correspond to any point in the associated projective space.

Theorem: The images in $V_{m,n}$ of $P_m(k) \times [q]$ and $P_m(k) \times [r]$ under ϕ intersect if and only if $q = r$.
Proof: The "if" part is of course trivial. Now let $q = (a_i)$ and $r = (b_i)$ and suppose $(x_i a_j) = (y_i b_j)$ where (x_i) and (y_i) are points of $P_m(k)$. Then there is an element $t \neq 0$ in k such that $x_i a_j = t y_i b_j$ for all i, j. Suppose $y_k b_l \neq 0$. Then $x_k a_l \neq 0$ and so $x_k \neq 0$. Then $x_k a_j = t y_k b_j$ for $j = 0, 1, \cdots, n$. Therefore $a_j = t y_k x_k^{-1} b_j$ for $j = 0, 1, \cdots, n$. Now $t y_k x_k^{-1}$ is a non-zero element in k and so $(a_j) = (b_j)$, $q = r$. This proves the theorem.

It is obvious that all these considerations work similarly for mappings by ϕ of $[p] \times P_n(k)$ into $V_{m,n}$ where p is a fixed point of $P_m(k)$.

5. Quadrics as ruled surfaces

The results of the preceding section can be used to study the lines contained in a non-singular quadric surface in $P_3(k)$ where k is the field of complex numbers. We know that under these circumstances such a quadric is projectively equivalent to the Segre variety $V_{1,1}$ having the equation $z_{00} z_{11} = z_{01} z_{10}$. Every point on this quadric is of form $(x_0 y_0, x_0 y_1, x_1 y_0, x_1 y_1)$ where (x_0, x_1) and (y_0, y_1) range over all points of $P_1(k)$; the quadric consists precisely of these points. Now let $q = (a_0, a_1)$ and consider the image of $P_1(k) \times [q]$ in $V_{1,1}$. This image consists exactly of the points $(x_0 a_0, x_0 a_1, x_1 a_0, x_1 a_1)$. This set of points constitutes a straight line contained *in the surface* $V_{1,1}$. In particular $V_{1,1}$ contains the two points $(a_0, a_1, 0, 0)$ and $(0, 0, a_0, a_1)$, and the set of all points of the image consists of all linear combinations (omitting the trivial one—all terms zero) of these two. Thus the image consists of a straight line, which we shall denote by L_q. Clearly every point of the quadric lies on some such line line for some q, and by the theorem of §5, on exactly one such line. A surface generated by lines in this sense is called a *ruled surface* and we have shown that every complex projective non-singular quadric is such a ruled surface. It is easy to check that the equations of L_q as an algebraic set in $P_3(k)$ are given by the two equations

$$\begin{cases} a_1 z_{00} = a_0 z_{01}, \\ a_1 z_{10} = a_0 z_{11}. \end{cases}$$

It is left as an exercise to construct the similar family of lines in $V_{1,1}$ starting from the mapping $[p] \times P_1(k) \longrightarrow V_{1,1}$.

Exercises for Chapter 5

1. Show that if U is an algebraic set in $P_m(k)$ and V and algebraic set in $P_n(k)$ then $\phi(U \times V)$ is an algebraic subset of the Segre variety $V_{m,n}$.
2. Prove the "triviality" noted after equations (2) to the effect that there are points of $P_N(k)$ which do not satisfy them.

3. Prove that the change of variables used in the examples at the end of §2, and of §3, does indeed amount, in each case, to a projective transformation.
4. Verify that the equations for L_q in §5 are those given.
5. Carry out the suggestion at the end of §5 concerning the construction from $[p] \times P_1(k)$.
6. Describe and interpret pictorially the results of §5 for the hyperboloid of one sheet in real euclidean 3-space. (Hint: construct a projective transformation so that the homogeneous equation of the hyperboloid takes the form of that of $V_{1,1}$.)

VI

PLÜCKER CO-ORDINATES AND GRASSMANN VARIETIES

1. Parametrizing by algebraic sets

An important class of problems in algebraic geometry is concerned with parametrizing mathematical systems by means of algebraic sets. In its most general formulation it amounts to this: to construct for any mathematical system S, an algebraic set V in $A_n(k)$ or $P_n(k)$ such that the elements of S are in one-to-one correspondence with the points of V. Then V is said to *parametrize* S. This problem appears throughout the subject in all degrees from the completely trivial to the immensely difficult—and unsolved.

Let us consider a trivial example first. Consider the set S of lines in $A_2(k)$ which go through the point $(0, 0)$. All such lines are given by equations of the form $\lambda x_1 + \mu x_2 = 0$ where λ and μ are elements of k not both zero. Two such lines coincide if and only if the corresponding ordered pairs (λ, μ) are proportional This is true for any field k. Thus if the ordered pair (λ, μ) is interpreted as a point of $P_1(k)$, we have constructed a one-to-one correspondence between the lines of the system S and the algebraic set $P_1(k)$.

We now turn our attention to an example much less trivial, but still capable of solution by elementary methods.

2. Subspaces of projective spaces

Let $P_n(k)$ denote as usual the n-dimensional projective space over a field k. By a *subspace* of $P_n(k)$ is meant a subset M of $P_n(k)$ such that if p and q are in M, then the line joining them lies entirely in M. The linearity condition means that if $(x_0, \cdots, x_n)$ and $(y_0, \cdots, y_n)$ are distinct points of M and if λ and μ are elements of k not both zero, then $(\lambda x_0 + \mu y_0, \cdots, \lambda x_n + \mu y_n)$ is a point of M. All this has a very simple interpretation in terms of vector spaces. Suppose $P_n(k)$ is realized as the set of one-dimensional subspaces of the vector space V_{n+1} of $(n+1)$-tuples over k. Then a subspace M of $P_n(k)$ consists of the set of one-dimensional subspaces in a vector-subspace of V_{n+1}. If this vector subspace W_{m+1} is of dimension $m+1$ we say that the projective subspace M is of dimension m. If $w_0, \cdots, w_m$ is a basis for W_{m+1} the mapping

$$\begin{cases} w_0 \longrightarrow (1, 0, \cdots\cdots, 0), \\ w_1 \longrightarrow (0, 1, 0, \cdots, 0), \\ \cdot \\ \cdot \\ \cdot \\ w_m \longrightarrow (0, \cdots\cdots, 0, 1) \end{cases}$$

where the images are $(m+1)$-tuples, has a unique extension to a linear isomorphism of W_{m+1} with V_{m+1}, the vector space of $(m+1)$-tuples over k. Thus this linear isomorphism induces a one-to-one correspondence of M with $P_m(k)$ which is linear in the obvious sense. It is very important to realize, however, that this isomorphism must be constructed differently for each m-dimensional subspace of $P_n(k)$ since there is a profusion of such m-dimensional subspaces in $P_n(k)$ and they are related in very complicated ways.

We are now in a position to state the main problem: to find an algebraic set which parametrizes the set of m-dimensional subspaces of $P_n(k)$. This is equivalent, obviously, to parametrizing the set of $(m+1)$-dimensional vector subspaces of V_{n+1}. Now, it is known that any finite-dimensional vector space over k is isomorphic to V_{n+1} for some value of n. Indeed, if $e_0, \cdots, e_n$ is a basis for such a vector space, then the mapping

$$\begin{cases} e_0 \longrightarrow (1, 0, \cdots\cdots, 0), \\ e_2 \longrightarrow (0, 1, 0, \cdots, 0), \\ \cdot \\ \cdot \\ \cdot \\ e_n \longrightarrow (0, \cdots\cdots, 0, 1) \end{cases}$$

can be extended uniquely to a linear isomorphism of the given vector space with V_{n+1}. In this sense, the algebraic set we shall construct will parametrize the subspaces of given dimension of an arbitrary (abstract) finite-dimensional vector space over k. The algebraic set to be constructed is called the *Grassmann variety*. In what follows, we shall keep to the notation that has been established up to this point.

3. Plücker co-ordinates

Let $w_0, w_1, \cdots, w_m$ be a basis for W_{m+1} over k and suppose $w_i = (x_0{}^{(i)}, \cdots, x_n{}^{(i)})$, $i = 0, 1, \cdots, m$. Set

$$\pi_{i_0 i_1 \cdots i_m} = \begin{vmatrix} x_{i_0}^{(0)} & x_{i_1}^{(0)} & \cdots & x_{i_m}^{(0)} \\ \cdot & \cdot & & \cdot \\ \cdot & \cdot & & \cdot \\ \cdot & \cdot & & \cdot \\ x_{i_0}^{(m)} & x_{i_1}^{(m)} & & x_{i_m}^{(m)} \end{vmatrix}$$

where each index i_l ranges from 0 to n. We are going to show, first of all, that the quantities $\pi_{i_0 i_1 \cdots i_m}$ can be used as co-ordinates of a point in projective space of suitably high dimension and that this point depends only on the subspace W_{m+1}.

Not all the $\pi_{i_0 i_1 \cdots i_m}$ are zero since the vectors $w_0, \cdots, w_m$ are linearly independent. Also the interchange of any two indices changes the sign of $\pi_{i_0 \cdots i_m}$, and if two indices are equal, then $\pi_{i_0 \cdots i_m} = 0$. Thus these quantities are uniquely determined if we know those with $i_0 < i_1 < \cdots < i_m$. Now suppose $w_0', \cdots, w_m'$ is another basis of W_{m+1} over k where $w_i' = (y_0^{(i)}, \cdots, y_n^{(i)})$; $i = 0, 1, \cdots, m$. Then

$$(y_0^{(i)}, \cdots, y_n^{(i)}) = \sum_{j=0}^{m} a_{ij}(x_0^{(j)}, \cdots, x_n^{(j)})$$

where $a_{ij} \in k$ with $\det(a_{ij}) \neq 0$ and $i = 0, 1, \cdots, m$. Thus $y_l^{(i)} = \sum_{j=0}^{m} a_{ij} x_l^{(j)}$ for $l = 0, 1, \cdots, n$. Now set

$$\pi'_{i_0 i_1 \cdots i_m} = \begin{vmatrix} y_{i_0}^{(0)} & y_{i_1}^{(0)} & \cdots & y_{i_m}^{(0)} \\ \cdot & \cdot & & \cdot \\ \cdot & \cdot & & \cdot \\ \cdot & \cdot & & \cdot \\ y_{i_0}^{(m)} & y_{i_1}^{(m)} & \cdots & y_{i_m}^{(m)} \end{vmatrix}$$

where each index i_l ranges from 0 to n. By the rule for multiplying determinants it follows that $\pi'_{i_0 i_1 \cdots i_m} = \delta \pi_{i_0 \cdots i_m}$ where $\delta = \det(a_{ij}) \neq 0$. Thus if the quantities $\pi_{i_0 i_1 \cdots i_m}$ are regarded as homogeneous co-ordinates of a point in projective space of suitable dimension, then this point *depends only on the subspace* W_{n+1}, and not on the choice of basis.

It is now high time to discuss the projective space for which the quantities $\pi_{i_0 \cdots i_m}$ are homogeneous co-ordinates.

As was pointed out before, it is essential to know only those $\pi_{i_0 i_1 \cdots i_m}$ for which $i_0 < i_1 < \cdots < i_m$. The number of these "essential" co-ordinates is therefore $\binom{n+1}{m+1}$. Set $N = \binom{n+1}{m+1} - 1$. We then regard the quantities $\pi_{i_0 i_1 \cdots i_m}$ with $i_0 < i_1 < \cdots < i_m$ as homogeneous co-ordinates of a point in $P_N(k)$ and denote this point by $(\pi_{i_0 \cdots i_m})$.

What we have constructed so far is a well-defined mapping ψ of the set of $(m+1)$-dimensional subspaces of V_{n+1} into the projective space $P_N(k)$. The quantities $\pi_{i_0 \cdots i_m}$ are called the *Plücker co-ordinates* of W_{m+1} (or of M). Those with $i_0 < i_1 < \cdots < i_m$ we shall call *reduced Plücker co-ordinates*. [Note: this concept of *reduced* Plücker co-ordinates is not standard in the literature; it is used here only in the interests of clarity.]

The next step is to show that ψ is a *one-to-one* mapping of the set of $(m+1)$-dimensional subspaces of V_{n+1} into $P_N(k)$; in other words, the point $(\pi_{i_0 i_1 \cdots i_m})$ of $P_N(k)$ determines W_{n+1} uniquely. We begin with the fact that a vector $(\lambda_0, \cdots, \lambda_n) \in V_{n+1}$ belongs to the subspace W_{m+1} if and only if the matrix

$$\begin{pmatrix} \lambda_0 & \lambda_1 & \cdots & \lambda_n \\ x_0^{(0)} & x_1^{(0)} & \cdots & x_n^{(0)} \\ \cdot & \cdot & & \cdot \\ \cdot & \cdot & & \cdot \\ \cdot & \cdot & & \cdot \\ x_0^{(m)} & x_1^{(m)} & \cdots & x_n^{(m)} \end{pmatrix}$$

has rank $m+1$. The vectors $w_0, \cdots, w_m$ are linearly independent and so the rank is at least $m+1$. Consequently the requirement that the rank be exactly $m+1$ means that the vector $(\lambda_0, \cdots, \lambda_n)$ must be a linear combination of $w_0, \cdots, w_m$. An arbitrary sub-determinant of degree $m+2$ has the form

$$\begin{vmatrix} \lambda_{i_0} & \cdots & \lambda_{i_m} & \lambda_{i_{m+1}} \\ x_{i_0}^{(0)} & \cdots & x_{i_m}^{(0)} & x_{i_{m+1}}^{(0)} \\ \cdot & & \cdot & \cdot \\ \cdot & & \cdot & \cdot \\ \cdot & & \cdot & \cdot \\ x_{i_0}^{(m)} & \cdots & x_{i_m}^{(m)} & x_{i_{m+1}}^{(m)} \end{vmatrix}$$

where the indices i_l range from 0 to n. The requirement on the rank of the matrix means that all these determinants must vanish. Expanding such a determinant by minors of the first row, we obtain the equation

$$(1) \qquad \sum_{k=0}^{m+1} (-1)^k \lambda_{i_k} \pi_{i_0 \cdots i_{k-1} i_{k+1} \cdots i_{m+1}} = 0.$$

We get such an equation for every choice of the index set $i_0, i_1, \cdots, i_m, i_{m+1}$. Clearly if the equations are satisfied in the case $i_0 < i_1 < \cdots < i_m < i_{m+1}$ then they are also satisfied in the arbitrary case and so we need only consider those with the indices in increasing order. The number of these "essential" equations is thus $\binom{n+1}{m+2}$. They express a necessary and sufficient condition in terms of the point $(\pi_{i_0 \cdots i_m})$ for a vector to belong to W_{m+1}. Thus the Plücker co-ordinates do indeed determine the subspace W_{m+1} uniquely. The equations (1) may be interpreted in terms of either the reduced Plücker co-ordinates, in which case there are $\binom{n+1}{m+2}$ equations, or else in terms of arbitrary Plücker co-ordinates, in which case there are many more equations but not any which are not equivalent to one in "reduced" form.

4. Grassmann varieties

To recapitulate, we now have a mapping ψ which maps the set S of $(m+1)$-dimensional subspaces of V_{n+1} into $P_N(k)$.

We have just shown that ψ is one-to-one. The question now arises: is it surjective? Answer: it is not. We are going to show that $\psi(S)$ is contained in a proper algebraic subset of $P_N(k)$.

Let $\pi_{j_0 j_1 \cdots j_m}$ be the Plücker co-ordinates of W_{m+1}. We shall not assume they are reduced; in other words, we take all of them. Now, for fixed values of $j_0, j_1, \cdots, j_{m-1}$ consider the $(n+1)$-tuple

$$v = (\pi_{j_0 j_1 \cdots j_{m-1} 0}, \pi_{j_0 j_1 \cdots j_{m-1} 1}, \cdots, \pi_{j_0 j_1 \cdots j_{m-1} n})$$

as a vector in V_{n+1}. We are going to show that v is contained in W_{m+1}. A vector $(y_0, \cdots, y_n)$ is in W_{m+1} if and only if there are elements $c_0, \cdots, c_m$ in k such that

$$(y_0, \cdots, y_n) = \sum_{i=0}^{m} c_i (x_0^{(i)}, \cdots, x_n^{(i)}),$$

that is, if and only if $y_j = \sum_{i=0}^{m} c_i x_j^{(i)}$ for $j = 0, \cdots, n$.

Now

$$\pi_{j_0 j_1 \cdots j_m} = \begin{vmatrix} x_{j_0}^{(0)} & x_{j_1}^{(0)} & \cdots & x_{j_m}^{(0)} \\ \cdot & \cdot & & \cdot \\ \cdot & \cdot & & \cdot \\ \cdot & \cdot & & \cdot \\ x_{j_0}^{(m)} & x_{j_1}^{(m)} & \cdots & x_{j_m}^{(m)} \end{vmatrix}.$$

We expand this determinant by the minors of the last column to conclude that

$$\pi_{j_0 j_1 \cdots j_m} = \sum_{i=0}^{m} c_i x_{j_m}^{(i)}$$

where $c_0, \cdots, c_m$ are elements in k. Thus it follows that $v \in W_{m+1}$.

From this we conclude immediately that the co-ordinates of v must satisfy the equations (1). (For convenience, since we are using non-reduced Plücker co-ordinates, we shall count all the "redundant" equations also.) Making the substitution of $\pi_{j_0 j_1, \cdots, j_{m-1} i_k}$ for λ_{i_k} in equations (1), we obtain

$$(2) \quad \sum_{k=0}^{m+1} (-1)^k \pi_{j_0 j_1, \cdots j_{m-1} i_k} \pi_{i_0 \cdots i_{k-1} i_{k+1} \cdots i_{m+1}} = 0.$$

Even though these equations are presented in terms of non-reduced co-ordinates, it is apparent immediately that we need only consider those equations with $i_0 < i_1 < \cdots < i_{m+1}$ and $j_0 < \cdots < j_{m-1}$. Not all authors write these equations in the same way. In van der Waerden [14] they are given in the form

$$(2') \quad \pi_{i_0 \cdots i_m} \pi_{j_0 \cdots j_m} = \sum_{k=0}^{m} \pi_{j_k i_1 \cdots i_m} \pi_{j_0 \cdots j_{k-1} i_0 j_{k+1} \cdots j_m}.$$

The deduction of (2′) from (2) is left as an exercise. [See Exercise 1 where a hint is given.]

We now introduce independent variables $p_{i_0 \cdots i_m}$ where the indices i_l range from 0 to n. On these variables we impose the conditions

$$(3) \quad \begin{cases} p_{i_0 \cdots i_m} p_{j_0 \cdots j_m} = \sum_{k=0}^{m} p_{j_k i_1 \cdots i_m} p_{j_0 \cdots j_{k-1} i_0 j_{k+1} \cdots j_m}; \\ p_{i_0 \cdots i_m} = 0 \text{ if two indices are the same;} \\ p_{i_0 \cdots i_m} \text{ reverses its sign if two indices are interchanged.} \end{cases}$$

For any assignment of values to these variables which satisfies the conditions (3), the quantities $p_{i_0 i_1 \cdots i_m}$ with $i_0 < i_1 < \cdots < i_m$ are homogeneous co-ordinates of a point in $P_N(k)$ provided these quantities are not all zero. The conditions (3) amount to a set of homogeneous polynomial equations and so they determine an algebraic set G_m in $P_N(k)$. This algebraic set is called the *Grassmann variety** associated with the set of $(m + 1)$-dimensional subspaces of V_{n+1}. G_m consists of the points $(p_{i_0 i_1 \cdots i_m})$, $i_0 < i_1 < \cdots < i_m$, satisfying the conditions (3).

Remark: From one point of view, it would be desirable to rephrase the first equation of (3) entirely in terms of *reduced* Plücker co-ordinates, in which case there would be no need to introduce the variables $p_{i_0 i_1 \cdots i_m}$ where the indices are not in natural order. We have not done this, since it does not coincide with the practice in the standard literature. Another way out of the inconvenience is simply to use *all* the variables $p_{i_0 i_1 \cdots i_m}$ as homogeneous co-ordinates in a projective space of dimension $(n + 1)^{m+1} - 1$.

In general G_m is a *proper* algebraic subset of $P_N(k)$. In case $m = n$, the only solution to the equations (3) is $p_{01 \cdots n} = 1$. This reflects the fact that V_{n+1} has only one subspace of dimension $n + 1$, namely the whole space. In this case $N = \binom{n+1}{m+1} - 1 = 0$ and so in this trivial case $G_m = P_0(k)$. For the case of two-dimensional subspaces in V_4, or of one-dimensional projective subspaces in $P_3(k)$, the conditions (3) reduce to the single equation

$$(4) \qquad p_{01}p_{23} - p_{02}p_{13} + p_{03}p_{12} = 0$$

for the reduced Plücker co-ordinates. This we can regard

*G_m satisfies the technical requirement (absolute irreducibility) for the term "variety" to be applicable.

as a quadric in $P_5(k)$ since $N = \binom{4}{2} - 1 = 5$ and obviously it is a proper subset of $P_5(k)$.

The final step consists in showing that the mapping $\psi : S \longrightarrow G_m$ is surjective. In other words, the $(m + 1)$-dimensional subspaces of V_{n+1} are in one-to-one correspondence with the points of G_m, so that G_m parametrizes S. Unfortunately the general proof of this is rather nasty; the reader is referred to Hodge and Pedoe [7] for a complete proof. We shall prove it here ultimately only in the case $m = 1$. Suppose the quantities $p_{i_0 i_1, \cdots i_m}$ are evaluated to satisfy (3) and that at least one of them is not zero. By relabeling indices and multiplying all the quantities by an appropriate non-zero factor, we may assume that $p_{01\cdots m} = 1$. Now for $i = 0, 1, \cdots, m$ and $j = 0, \cdots, n$ we set $x_j^{(i)} = p_{0 \cdots i-1, j, i+1, \cdots m}$. Then $x_j^{(i)} = \delta_j^i$ for $0 \leq j \leq m$ where δ_j^i is the Kronecker delta ($\delta_j^i = 1$ if $i = j$ and $\delta_j^i = 0$ if $i \neq j$). Then the vectors

$$w_i = (x_0^{(i)}, x_1^{(i)}, \cdots, x_n^{(i)})$$

in V_{n+1} are linearly independent. The proof of surjectivity consists in showing that the Plücker co-ordinates of the subspace spanned by these vectors coincide, to within a non-zero factor in k, with the given quantities $p_{i_0 i_1 \cdots i_m}$.

In the case $m = 1$, we compute the Plücker co-ordinates from the vectors w_0 and w_1 to obtain

$$\pi_{0i} = \begin{vmatrix} x_0^{(0)} & x_i^{(0)} \\ x_0^{(1)} & x_i^{(1)} \end{vmatrix} = \begin{vmatrix} 1 & p_{i1} \\ 0 & p_{0i} \end{vmatrix} = p_{0i},$$

$$\pi_{1i} = \begin{vmatrix} x_1^{(0)} & x_i^{(0)} \\ x_1^{(1)} & x_i^{(1)} \end{vmatrix} = \begin{vmatrix} 0 & p_{i1} \\ 1 & p_{0i} \end{vmatrix} = -p_{i1} = p_{1i}$$

for $i = 0, 1, \cdots, n$. It is easy to check that the first equation of (3) can be written in the form

$$p_{gh}\, p_{kl} - p_{gk}\, p_{hl} + p_{gl}\, p_{hk} = 0, \tag{5}$$

where g, h, k range from 0 to n. Now, these equations are satisfied by the π_{gh} since they are Plücker co-ordinates of a two-dimensional subspace of V_{n+1}. In case $k = 0$, $l = 1$ the equations take the form $p_{gh} = p_{g0}\, p_{h1} - p_{g1}\, p_{h0}$. From the above computations, the right-hand side of this equation is equal to $\pi_{g0}\, \pi_{h1} - \pi_{g1}\, \pi_{h0}$ and so $p_{gh} = \pi_{gh}$ for all $g, h = 0, 1, \cdots, n$. Thus every point of G_1 corresponds to a two-dimensional vector subspace of V_{n+1} and we know from §3 that this correspondence is one-to-one. In the case $n = 3$, the equations (5) reduce to the single equation (4), so that a one-to-one correspondence has been constructed between lines in $P_3(k)$ and points of a certain quadric in $P_5(k)$.

5. Grassmann varieties as homogeneous spaces

An invertible linear transformation of V_{n+1} maps any vector subspace into another of the same dimension, and so induces a (point) mapping of G_m into itself for each $m = 0, 1, \cdots, n$. Furthermore, given any two $(m + 1)$-dimensional subspaces of V_{n+1}, there always exists an invertible linear transformation mapping the first onto the second. The set of invertible transformations of V_{n+1} constitute the *full linear group* $GL_{n+1}(k)$. Thus, this group operates on G_m — in the sense that it can be construed as a group of transformations of G_m. Furthermore, the operation of $GL_{n+1}(k)$ on G_m is *transitive* in the sense that if p and q are any two points of G_m then there is an element Φ of $GL_{n+1}(k)$ such that $\Phi(p) = q$. This situation is expressed by saying that G_m is a *homogeneous space* for $GL_{n+1}(k)$.

To define the notion in its purest algebraic form, suppose that G is an arbitrary group. Then a *homogeneous space* for G is any set S on which G acts transitively as a group of transformations. To take a simple but important example, let G be any group and H an arbitrary subgroup of G. If S is the set of left-cosets $a \cdot H$, $a \in G$, then G operates transitively on S by means of transformations of the form $\phi_g : a \cdot H \longrightarrow (g \cdot a) \cdot H$, $g \in G$, and so S is a homogeneous space for G. In most situations where homogeneous spaces arise, both the group G and the set S have some additional structure; for instance they may be algebraic sets or topological spaces. In such a case it is required that G operate on S in a manner which is admissible in the sense of the extra structure.

In the case of $GL_{n+1}(k)$ operating on G_m suppose that $\Phi(p) = q$ when $p, q \in G_m$ and $\Phi \in GL_{n+1}(k)$. Then the co-ordinates of q are homogeneous polynomials in the co-ordinates of p, and with coefficients which are polynomials in the co-ordinates of Φ. It is understood here, of course, that $GL_{n+1}(k)$ is embedded as an algebraic set in $A_{(n+1)^2+1}(k)$ as in Chapter IV.

Exercises for Chapter VI

1. Reduce equations (2) to equations (2′). Hint: make the following replacement of indices:

$$\begin{array}{ccccccc} j_0 & \cdots & j_{m-1} & i_0 & i_1 & \cdots & i_{m+1} \\ \downarrow & & \downarrow & \downarrow & \downarrow & & \downarrow \\ i_1 & \cdots & i_m & i_0 & j_0 & \cdots & j_m \end{array}$$

2. Prove that the quadric given by equation (4) is non-singular.

BIBLIOGRAPHY

The following list contains not only books mentioned in the text but also several other standard references which have been consulted. The latter are included for the benefit of readers who wish to pursue the subject further.

[1] M. Baldassarri: *Algebraic Varieties*, Berlin, 1956.
[2] Birkhoff and MacLane: *A Survey of Modern Algebra*, New York, 1953.
[3] N. Bourbaki: *Éléments de mathématiques*, Paris, 1939 et seq.
[4] C. Chevalley: *Fundamental Concepts of Algebra*, New York, 1956.
[5] P. R. Halmos: *Naïve Set Theory*, Princeton, 1960.
[6] P. R. Halmos: *Finite-Dimensional Vector Spaces*, 2nd ed., Princeton, 1958.
[7] Hodge and Pedoe: *Methods of Algebraic Geometry*, 3 vols., Cambridge, 1952 - 54.
[8] Hoffman and Kunze: *Linear Algebra*, Englewood Cliffs, N. J., 1961.
[9] A. Jaeger: *Introduction to Analytic Geometry and Linear Algebra*, New York, 1960.
[10] S. Lang: *Introduction to Algebraic Geometry*, New York, 1958.
[11] S. Lefschetz: *Algebraic Geometry*, Princeton, 1953.
[12] T. Nagell: *Introduction to Number Theory*, New York, 1951.
[13] P. Samuel: *Méthodes d'algèbre abstraite en géométrie algébrique*, Berlin, 1955.
[14] B. L. van der Waerden: *Einführung in die algebraische Geometrie*, Berlin, 1939; New York, 1955.
[15] B. L. van der Waerden: *Moderne Algebra* (I), 5th ed., Berlin, 1960.
[16] R. Walker: *Algebraic Curves*, Princeton, 1950.
[17] A. Weil: *Foundations of Algebraic Geometry*, New York, 1946.
[18] Zariski and Samuel: *Commutative Algebra*, 2 vols., Princeton, 1958, 1960.

INDEX